American Institute of Chemical Engineers

Energy and Transport Processes

Presentations at the
2007 AIChE Annual Meeting

November 4-9, 2007
Salt Lake City, Utah, USA

Printed from e-media with permission by:

Curran Associates, Inc.
57 Morehouse Lane
Red Hook, NY 12571
www.proceedings.com

ISBN: 978-1-60560-005-5

Some format issues inherent in the e-media version may also appear in this print version.

ISBN: 978-1-60560-005-5

American Institute of Chemical Engineers
Proceedings
Three Park Avenue
New York, NY 10016-5991
Phone: 212-591-8100

www.aiche.org

American Institute of Chemical Engineers

Energy and Transport Processes

TABLE OF CONTENTS

268a Reaction Dynamics in a Parallel Flow Channel Pem Fuel Cell

Jay B. Benziger, Erin Kimball, E.-S. J. Chia, and Yannis G. Kevrekidis. Department of Chemical Engineering, Princeton University, Princeton, NJ 08544

The spatiotemporal dynamic response of a segmented anode parallel channel (SAPC) polymer electrolyte membrane (PEM) fuel cell was monitored following changes in flow rate, temperature and load resistance. Autohumidified operation with dry feeds at 1 bar pressure was achieved at temperatures below 70ºC, where the convective transport of water vapor was less than the water production by the fuel cell current. The current could be "ignited" by a single injection of water into the anode feed, or by reducing the temperature and external load resistance. Co-current flow of the hydrogen and oxygen resulted in current ignition at the outlets of the flow channels, followed by a wave of high current density propagating toward the inlets. Counter-current flow of the hydrogen and the oxygen resulted in ignition near the center of the flow channels; over time the ignition front fanned out. The spatio-temporal dynamics of the current ignition along the flow channels can be effectively predicted from a model of a set of coupled differential fuel cells in series. Liquid water condensing in the flow channels gives rise to complex spatio-temporal variations in the current density; these variations are strongly dependent on orientation of the fuel cell with respect to gravity.

268b Thermo-Mechanical Coupling in Particle Beds

Watson L. Vargas, University of Pittsburgh, 1249 Benedum Hall, University of Pittsburgh, Pittsburgh, PA 15261 and Joseph J. McCarthy, Chemical and Petroleum Engineering, University of Pittsburgh, 1238 Benedum Hall, Pittsburgh, PA 15261.

The consolidation of grains plays an important role in industries ranging from food, polymers, pharmaceuticals to ceramics and even in the development of novel nuclear reactor concepts. In this work, transient heat conduction in a quasi-static bed of granular material is studied computationally, both in 2D and 3D, using an extension of -- Thermal Particle Dynamics (TPD) -- a technique developed by the authors [Vargas and McCarthy, AIChEJ, 2001]. Specifically, the TPD technique, which yields detailed mechanical and thermal information not easily measured in experiments, is modified to include the effects of thermo-mechanical coupling. We use this modified technique to explore the evolution of the force distribution and void fraction when a bed of grains is subjected to thermal cycling. We show that our results are in agreement with recent experimental findings [K. Chen {\it et al.} , Nature 2006]. That is, we see that a controlled increase in granular densification can be induced by thermal cycling, without resorting to the input of mechanical energy. This thermo-mechanical behavior may have important practical implications for the storage and handling of granular materials.

268c Dissolution Behavior of Coated Microbubbles and Fluorescence Microscope of Glycerides Monolayer at Air/water Interface

Yuyi Shen, Chemical Engineering, University of California, Davis, One Sheilds Ave, Davis, CA 95616, Robert L. Powell, Chemical Engineering and Material Science, University of California-Davis, Davis, CA 95616, and Marjorie L. Longo, Chemical Engineering and Materials Science, University of California, Davis, Davis, CA 95616.

A low cost food grade emulsifier (a mixture of monoglycerides, diglycerides and sodium stearoyl lactylate) in combination with polyethylene glycol (PEG)-40 stearate was used as an alternative to pure staturated phospholipids to form the thin shell of a microbubble (micron-scale hollow spheres (normally 10–200 um) with a gaseous core coated with a thin shell). Polydisperse coated microbubbles can be generated using a probe sonification technique. To investigate the stability of microbubble in water system over time, shell emulsifier monolayer isotherm with fluorescence microscope was used. We present that shell microstructure of the mixed monolayer on air/water interface was changed by varying the concentration ratio of emulsifiers (food grade emulsifier and PEG-40s) concentration. The monolayer microstructure correlated the surface morphology of microbubbles dissolution behavior under degassed media with various degassing degree. Those dissolution behavior involving the buckling and rupture of the coated monolayer can be explained by the phase behavior of emulsifiers and their role in stability of microbubbles in water systems.

268d Condensation of Organic Vapors on Finned Tubes

Subrahmaniyam Susarla, CHEMICAL ENGINEERING, OSMANIA UNIVERSITY,UNIVERSITY COLLEGE OF TECHNOLOGY, -, -, HYDERABAD-500007, India and Shyamsunder Rachamalla, Technology, OSMANIA UNIVERSITY,UNIVERSITY COLLEGE OF TECHNOLOGY, -, -, HYDERABAD-500007, India.

The heat transfer coefficients of condensing organic vapors on horizontal/vertical tubes are much lower than that of condensing steam. Literature shows increasing condensation rates on finned tubes. Experiments were conducted with organic vapors condensing on a fabricated aluminum finned tube. The test condenser was an aluminum tube 300mm long, 24 mm OD and 21mm ID on which four longitudinal fins (10mm height and 1.2 mm thickness) were attached at equal angle circumferentially. Condensation of water, methanol, carbon tetrachloride and n-Hexane were carried at near atmospheric pressure on horizontal orientation of condenser. The details of experimental setup, measurements and instrumentation would be given. A preliminary data analysis indicated that the Nusselt analysis is not satisfactory in explaining the data. A video graphic study was carried out to check the nature of condensation. It was found that the condensation of organic vapors occurred in mixed mode i.e., both film wise and drop wise. Further data are being obtained and the data analysis would include:

•vapor entry velocity effect on the heat transfer coefficient •Effect of coolant thermal resistance variation along the condenser •Contribution of drop wise condensation to overall heat transfer rate.

Multiscale Drying and Stress-Crack Formation in Corn Kernels

Pawan Singh Takhar and Jyoti Hundal. Animal and Food Sciences, Texas Tech University, International Center for Food Industry Excellence, Box 42141, Lubbock, TX 79409

Abstract:
The fluid transport process in the biopolymeric systems exhibit multiscale characteristics. These systems are mainly composed of solid polymeric matrix, vicinal (solvated) fluids and bulk fluids. Transport of fluid is influenced by the relaxation of the surrounding polymers. A generalized integro-differential fluid transport equation, which accounts for the effect of material relaxation on the fluid flow has been developed. Finite element method is being implemented using Comsol Multiphysics for developing the solution scheme for coupling the developed multiscale moisture transport equation with the thermomechanical stress equation. The method is being used for predicting stress-cracking during drying of corn kernels. The required model parameters are moisture diffusivity and stress relaxation function. The dual porosities (high and low permeability channels) of corn kernels are being accounted in the solution scheme. The methodology is flexible and is capable of implementing on biopolymeric systems with other geometric shapes. The model can incorporate heterogeneous, time varying material properties in different domains within the simulation process and accounts for the effect of glass transition on fluid transport.

Extended Summary:

Food grains are hetrobiopolymers, as they are abundant in starches, proteins and lipids. Polymers exhibit viscoelastic behavior when subjected to heat and pressure. On the application of heat, the molecular motion increases and the polymer becomes soft and flexible (Slade and Levine, 1991). Fluid transport processes are influenced by the relaxation of the surrounding polymers. A generalized integro-differential fluid transport equation, which can account the effect of material relaxation on the fluid flow, has been developed by Singh et al (2003).

The structure of corn is complex consisting of three distinct regions with varying material properties. Viscoelastic characteristics of a biopolymer alter the transport properties. In order to optimize the processing conditions, thermo mechanical characteristics of a viscoelastic material are incorporated at the multiscale level. To capture the dynamics of the crack formation, an in-depth approach is required, that aims at exploring the complex physico mechanical interactions that occur inside the grain when subjected to changing moisture and thermal conditions.

The glass transition theory is one of the concepts that have been proposed to explain the stress cracking during drying. According to this concept, the polymers exist in the glassy and rubbery states. Temperature elevation during drying leads to the onset of internal phase transitions in the polymers from glassy to the rubbery states. The resultant effect of internal transitions on the transport properties within the transition range are not very clear, owing to the complex nature of the biopolymers, only general research theories have been proposed to explain the overall effect. A multiscale approach is proposed which would enable to study the dynamics of stress development at micro, meso and macroscale during moisture transport. Finite element method is being implemented using Comsol Multiphysics for developing the solution scheme for coupling the developed multiscale moisture transport equation with the thermomechanical stress equation.

The required model parameters are moisture diffusivity and stress relaxation function. The dual porosities (high and low permeability channels) of corn kernels are being accounted in the solution scheme. The transient variation in the material properties undergoing dehydration process above and below the glass transition and the continuous and intermittent processes are studied with the help of this simulation model.

Initially, 2D triangular elements were used for modeling the coupled integro-differential equation. The solution scheme was developed for corn geometry and is able to provide the information on moisture gradients and moisture movement as governed by the material diffusivity values. The model is capable of being modified to adapt itself to the time varying geometries, due to shrinkage and collapse of the internal structure. It is also possible to include heterogeneous, time varying material properties within different material domains during the simulation process. The solution also incorporates the effect of viscoelastic relaxation on the fluid transport. The methodology is flexible and is capable of being implemented on complex biopolymeric systems having different geometric shapes. The model can accommodate implicit and explicit functions of material properties and is able to closely simulate the results of coupled systems involving multiple transport processes, including viscoelastic relaxation and fluid – structure interactions during drying.

Acknowledgement:
Thanks to USDA-CSREES for providing financial support under the award number 2003-35503-13963.

References:

Ferry, J. 1980. *Viscoelastic properties of polymers*. NY: John Wiley and Sons Inc.

Irudayaraj, J., Haghighi, K. and Stroshine, R. L. 1992. Finite element analysis of drying with application to cereal grains. *Journal of Agricultural Engineering Research*, **53,** 209-229.

Irudyaraj, J. and Haghighi, K. 1993. Stress analysis of viscoelastic materials during drying. I. Theory and finite element formulation. *Drying Technology*, **11,** 901-927.

Singh, P. P., Cushman, J. H. and Maier, D. E. 2003. Multiscale fluid transport theory for swelling biopolymers. *Chemical Engineering Science*, **58,** 2409-2419.

Singh, P. P., Maier, D. E., Cushman, J. H., Haghighi, K. and Corvalan, C. 2004. Effect of viscoelastic relaxation on moisture transport in foods. Part I: Solution of general transport equation. *J. Math. Biol.*, **49,** 1-19.

Slade, L. and Levine, H. 1991. A food polymer science approach to structure-property relationships in aqueous food systems in Non-equilibrium behavior of carbohydrate-water systems. In: *Water Relationships in Foods* (Ed. by Levine, H., Slade, L.), pp. 805-823. NY: Plenum Publications.

Wannanen, K. M. and Okos, M. R. 1987. Viscoelastic Properties of Yellow-dent corn kernels. In: *ASAE Annual International Meeting, Chicago, IL*: ASAE.

268f Multiphase and Multicomponent Transport with Phase Change in Meat as Hygroscopic Porous Media

Ashish Dhall, Amit Halder, and Ashim K. Datta. Dept. of Biological and Environmental Engineering, Cornell University, 175 Riley Robb, Ithaca, NY 14850

Food borne diseases cause 76 million illnesses in the United States each year [1]. Since many of these diseases result from insufficient destruction of pathogens, safety of cooked meat is one of the primary concerns among food scientists and technologists. Safety of a food material includes protection from microbial, chemical and physical hazards or contamination that may occur at any stage of food production and handling. As far as meat cooking is concerned, time-temperature history determines microbial destruction and thus, is the primary indicator of safety. The cooking recommendations for safe for consumption meat are prescribed in terms of temperature and holding times of the slowest heating points [2, 3] to guard against lethal microbial activity. In addition to safety, the eating quality of the meat product, characterized by nutrient composition, flavor, color and texture/tenderness, is of importance to the consumer. It is reported that as meat is cooked, its toughness lasts till 50 °C, after which the meat product starts becoming tender and thus tasteful [4]. Elasticity of meat is adversely affected above 65 °C leading to a tougher, more brittle meat as one goes to higher temperatures. Thus, quality and safety put opposing demands on the temperature range and the most tasty and microbially safe meat requires cooking within a narrow temperature range. Apart from temperature, water and fat content are other two important components determining meat quality. Moisture content and distribution inside the meat protein fibers and extracellular spaces determine its juiciness, while flavor is mainly a property of fats and lipids. With more than 32% of its population as obese (Body Mass Index, BMI > 30) and another 34% overweight (25 < BMI < 30) [5], US needs improvements in the quality of its food products esp. in controlling fat content. Computer modeling of a meat cooking process help a long way in establishing a better understanding of the transport processes that take place inside which is vital for quality prediction and improvement. However, underlying physics is complex due to highly heterogeneous nature of the material.

In the current study, a macroscopic multiphase model based on unsaturated fluid flow in a hygroscopic porous media is developed, which accounts for all the important physical phenomena that take place while cooking of meat. Frozen meat is considered as a porous solid comprising of water, fat and protein with gas trapped in its pores. As it is heated, water and fat melt and are gradually released from the solid to the pore space according to their holding capacities in the solid matrix, which are assumed to be functions of temperature. As temperature further increases, free transportable water evaporates. Since four fluid components (liquid water, liquid fat, water vapor and air) are present in the pores, a mass balance equation is written for each component. Thermal equilibrium assumption leads to a single energy balance equation for the whole system. Melting of solid water and fat is handled by effective specific heat method. The other two phase changes are accounted for by adding source terms to the respective mass balance equations. The model is numerically solved in COMSOL Multiphysics and is validated for double sided contact heating of hamburger patties by comparing temperature and moisture profiles with experimental studies and numerical results available in literature. Dominant transport modes for water as well as fat under different conditions are identified by performing a sensitivity analysis on various parameters such as surface temperature, patty diameter, meat composition etc.

References

1. [cited May 8, 2007] http://www.cdc.gov/foodsafety/

2. Food Code. U.S. Public Health Service, Food and Drug Administration -- 1999 Washington, D.C.

3. USDA urges consumers to use food thermometer when cooking ground beef patties. FSIS News Release. Food Safety and Inspection Service. U.S. Department of Agriculture -- 1998 Washington, D.C.

4. Tornberg, E., Effects of heat on meat proteins - Implications on structure and quality of meat products. Meat Science, 2005. 70(3): p. 493-508.

5. [cited May 8 2007] http://win.niddk.nih.gov/statistics/index.htm.

CFD Simulation of Bubble Columns: modeling of non-uniform gas distribution at sparger

M. R. Rampure[a,b] *and V. V. Ranade*[a]
[a] *Industrial Flow Modeling Group, National Chemical Laboratory, Pune-411 008, INDIA*
[b] *Department of Chemical Engineering, Indian Institute of Technology Bombay, Powai, Mumbai-400076, INDIA*

Introduction

Most of the CFD (Computational Fluid Dynamics) simulations of bubble column reactors are based on the assumption that gas is uniformly distributed at the sparger. All the investigators adjust lift and drag coefficients to match overall gas hold-up and liquid circulation (see Joshi, 2001and references cited therein). The experimental data used to evaluate and fit these model parameters is usually obtained with sieve plate sparger. The sieve plate spargers are known to lead to non-uniform gas distribution. Only fraction of total numbers of holes of such a sieve plate sparger are active at any instant and this leads to non-uniform distribution of gas. The active region of the sparger varies with time leading to complex unsteady flow. Typical dynamic flow behavior on the sparger plate for the air water system is shown in Figure 1. It is obvious that it is essential to capture non-uniform distribution of gas at the sparger for realistic simulations of gas-liquid flow in bubble columns. Such an attempt is made here.

Figure1. Bubble swarms oscillating on the sparger plate at superficial gas velocity of 0.05cm/s

Experimental

In this work, a bubble column (air-water system) of 0.2 m diameter and 2 m height with the uniform-hole sparger (hole diameter 1.2 mm and F.A. of 1.13%) was used. The dispersed height to diameter ratio of 5 was maintained with the superficial gas velocities ranging from 0.05-0.40 m/s. Various experimental technique as conductivity meter, dual-tip conductivity probe and wall pressure fluctuations are used by wall mounting the sensors at various axial locations (as shown in Figure 2). The Wall pressure fluctuations were measured using dynamic pressure transducers (PCB Piezotronics Inc., USA) located at different axial locations from the sparger. The pressure signal was sampled at the rate of 400Hz for a period of 1400 sec and low pass filtered to 50 Hz signal. The power spectra obtained using the ANTS (Utikar et.

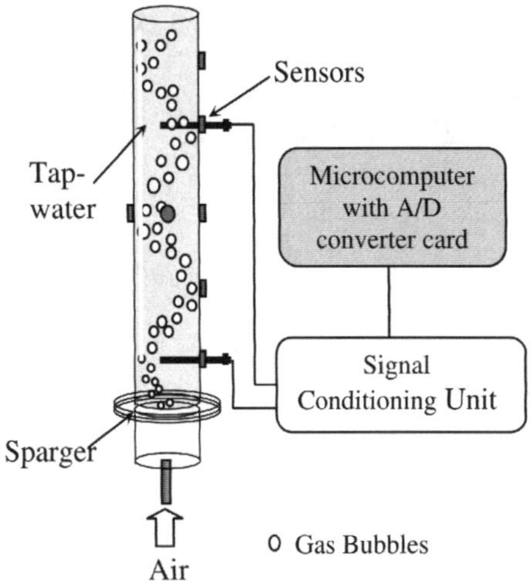

Figure 2. Experimental set-up

al., 2001) showed three major peaks at frequency of 0.2-0.3Hz, 3-5 Hz and 18-25Hz signifying bubble swarm oscillations, large bubble formation near sparger and movement of smaller bubbles respectively (Figure 3a). The power values for these maximum peaks when plotted against superficial gas velocity showed that the power values for 0.2-0.3Hz diminishes after 0.10m/s as more chaotic flow prevails. Also the slope changes for 18-25 Hz power value due to the cluster formation of the small bubbles (Figure 3b). The ratio of standard deviation to the mean of the pressure fluctuation series showed and change in the slope value at the superficial gas velocity of 0.02 m/s that continues upto 0.10m/s denoting the transition flow regime. Flow regime map is shown in Figure 3c.

Overall gas volume fraction was measured visually. Dual-tip conductivity probe was used to measure the local time averaged gas holdup, axial bubble velocity and bubble size distribution at two different axial locations (Rampure et. al., 2007). All the experiments were conducted at least three times. The experimental database was used to evaluate the CFD model.

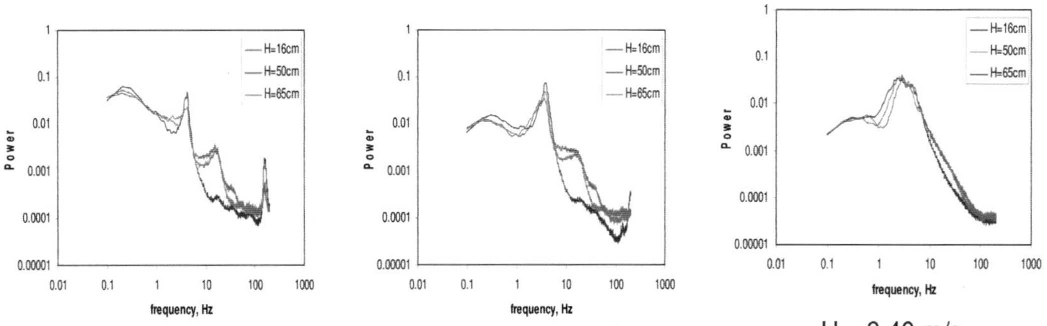

U_G=0.05 m/s U_G=0.10 m/s U_G=0.40 m/s

(a) Power spectral density

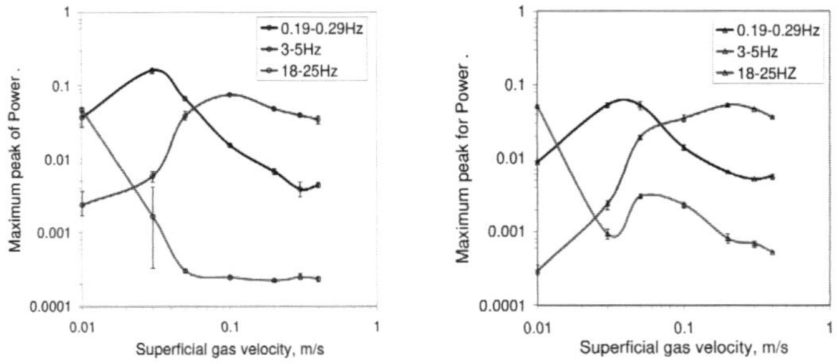

H=0.17m H=0.66m

(b) Maximum peak of power spectra

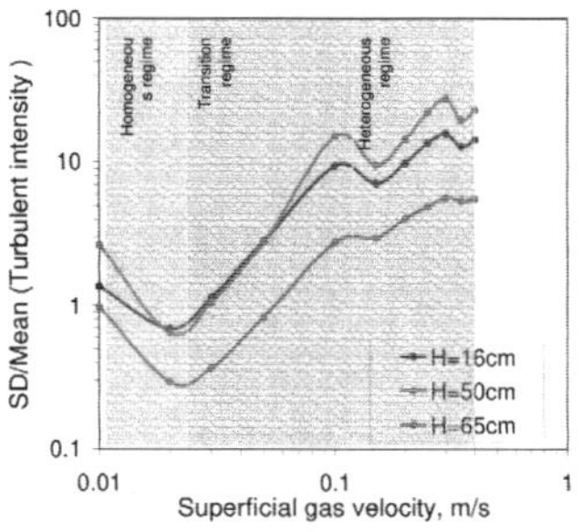

(c) Flow regimes

Figure 3. Wall pressure fluctuations

CFD model

More than a decade ago, Ranade (1993) have attempted to account for non-uniform distribution of gas and the sparger. However, the model considered axis-symmetric two-dimensional flow in bubble column and was a steady state model. In this work we have developed a comprehensive, three-dimensional, unsteady CFD model to simulate gas-liquid flow in bubble columns after accounting for non-uniform distribution of gas at the sparger. CFD model was developed using Eulerian approach for two-phase system. The standard k-ε mixture turbulence model was used to incorporate effect of turbulence. Correction of ε_L^2 (liquid holdup)to the drag law was used to consider effect of bubble swarm. Both the plenum section and the sparger plate above were considered in the solution domain for the first time (Figure 4). The uniform gas velocity was set as the inlet to the bottom of plenum section considering the volume fraction of gas and bubble rise velocity (Rampure et al., 2007). Sparger was modeled as porous zone. The viscous resistance and the inertial resistance for the perforated plate were estimated using the correlation suggested by Smith and Van-Winkle (1958). CFD simulations were carried out using commercial CFD code, FLUENT 6.3 (Ansys-Fluent Inc., USA). Typical snapshots of the unsteady CFD simulation are shown in Figure 5a-5b. Time averaging was done once quasi steady state achieved.

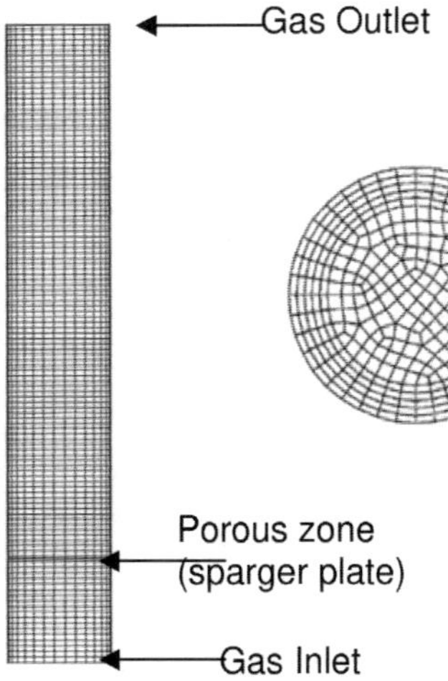

Figure 4 Computational domain

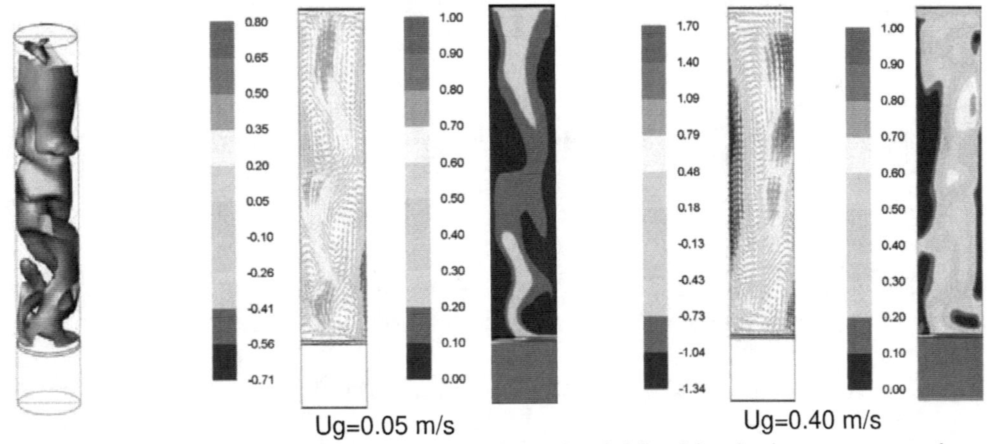

(a) Iso-surface of 0.2 gas holdup at Ug=0.20m/s

(b) Instantaneous snapshot of axial liquid velocity vectors and gas holdup contours

Figure 5. CFD simulations

Results

Non-uniform sparger implemented in the present work was compared with the uniform sparger boundary condition (Rampure et. al., 2007). It was observed that the overall gas holdup followed similar trend with experimental data. Divergence was detected for the drag correction of ε_L^2 while simulating for uniform sparger at Ug=0.4m/s, hence correction of ε_L^4 was used. But in the present study of non-uniform sparger drag correction of ε_L^2 was used for all four gas velocities (Figure 6a). Local time averaged gas holdup for non-uniform gas distribution at the sparger showed good agreement with the experimental data in comparison with uniform gas holdup at gas inlet boundary (Figure 6b).

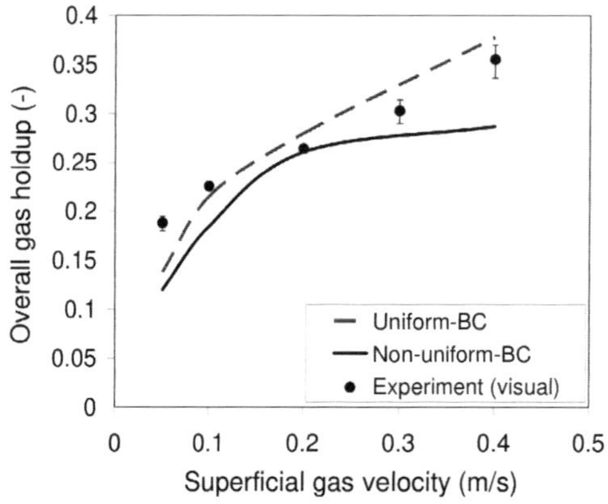

(a) Overall gas holdup

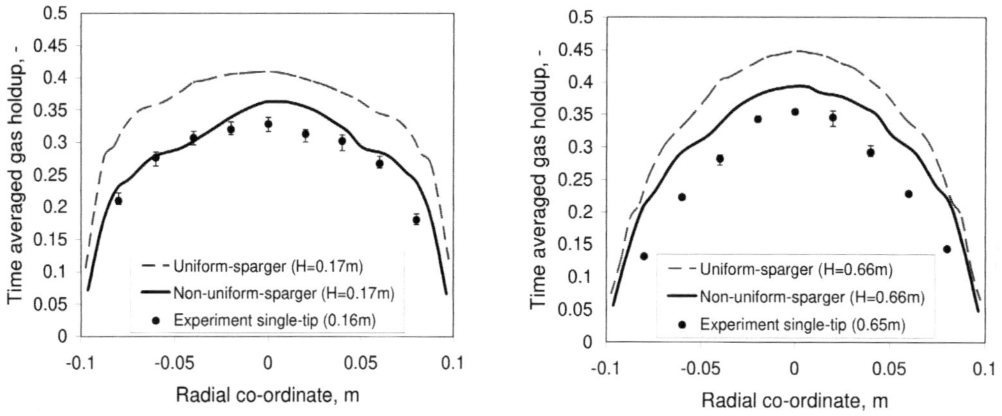

H=0.17 m H=0.66 m

(b) Time averaged local gas holdup at superficial gas velocity of 0.20 m/s

Figure 6. CFD results predicted for uniform & non-uniform gas distribution at sparger

Figure 7 shows the radial profiles for time-averaged local gas hold-up for three different gas velocities and at two different axial locations. The results predicted by the CFD model showed good agreement with the experimental data.

The profiles of time averaged axial liquid velocity obtained from the CFD simulations are shown in Figures 8a-8c. The CFD simulations showed good comparison with the experimental data of Sanyal et al. (using the CARPT technique) at superficial gas velocity of 0.10 m/s (Figure 8b). For gas velocity 0.40 m/s, not much difference in the values of time-averaged axial velocity profile was observed for prediction done at two axial locations (Figure 8c).

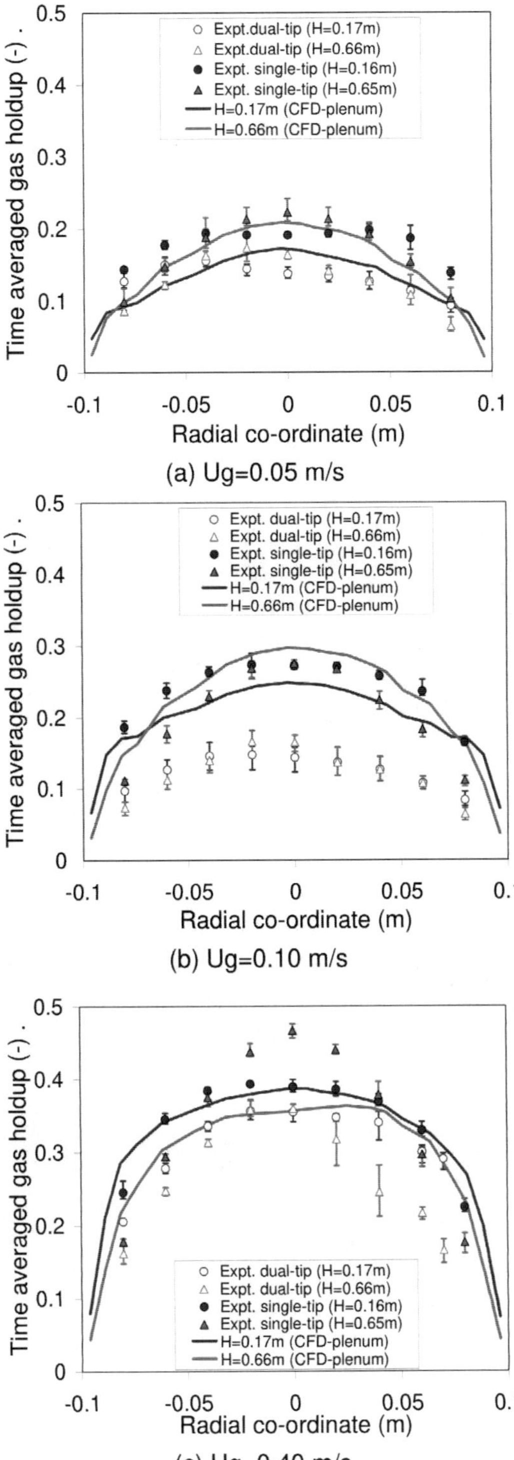

Figure 7. Time averaged gas holdup

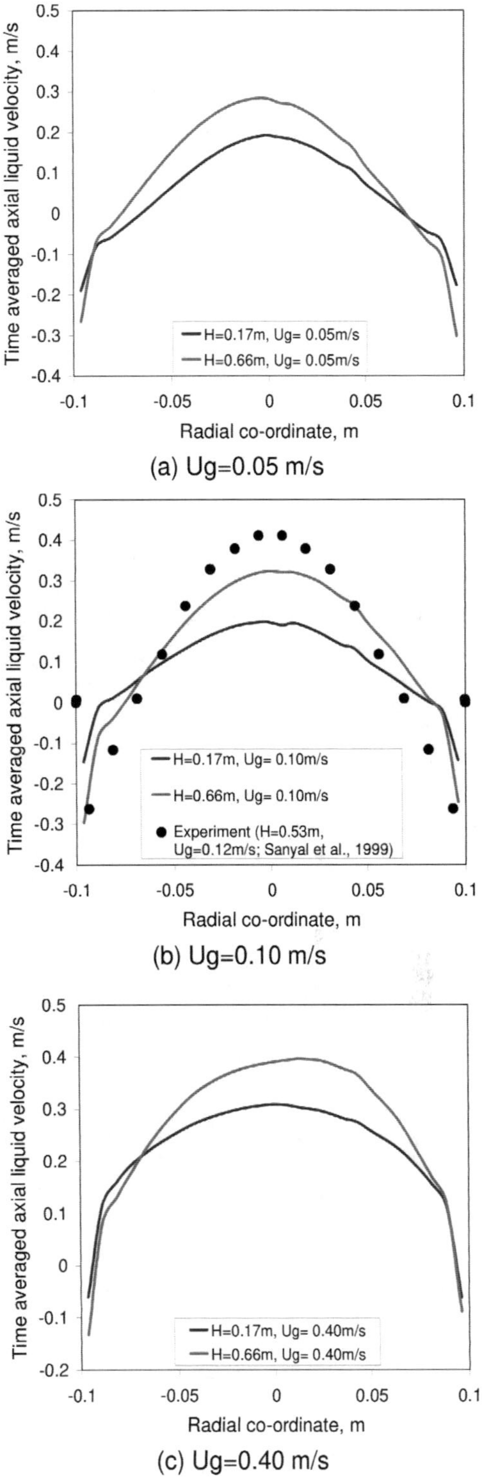

Figure 8. Time averaged axial liquid velocity

The measured and simulated radial profiles of time averaged axial gas velocity are shown in Figures 9. CFD simulations showed good agreement with the mean bubble velocity measured using the dual-tip conductivity probe (Figure 9a-9d). The comparison for higher superficial gas velocity however showed noticeable discrepancy (see Figure 9d). One of the reasons for this discrepancy may be the increased possibility of bubbles not cutting both the probe tips at higher gas velocity. This poses a difficulty in finding the pair of signals arising from the same bubble.

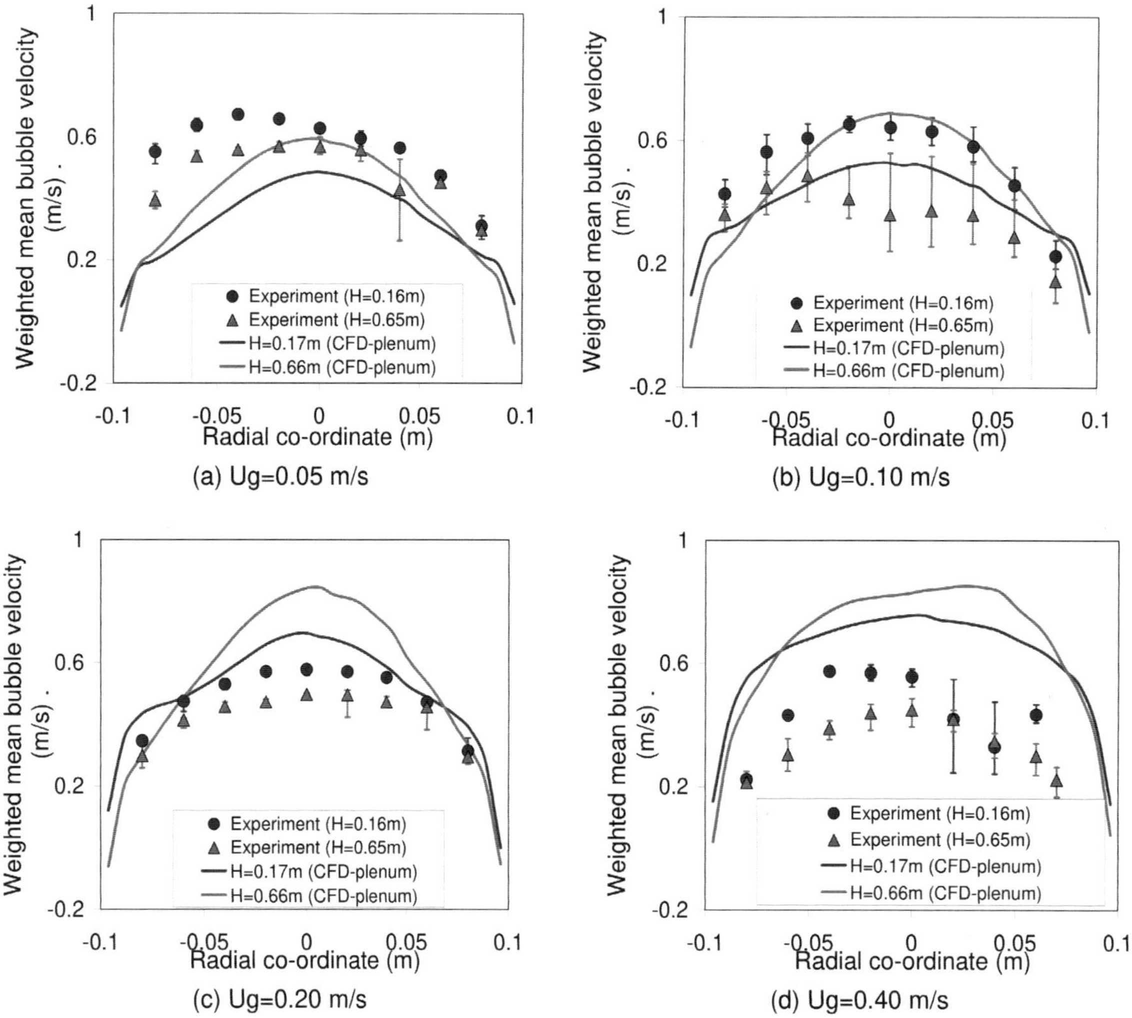

Figure 9. Time averaged axial gas velocity

The effect of resistances in the porous zone (sparger) was studied to estimate the changes in flow characteristics such as gas holdup, liquid velocity profiles and mixing time. Almost uniform distribution of the gas holdup on the sparger plate was observed for lower gas velocity compared to that at higher gas velocity of 0.2 m/s. The dynamic nature for flow on the sparger was more predominant for increase in gas velocity. Increase in sparger resistance (leading to more uniformity in gas distribution) for lower gas velocity lead to increase in the predicted gas holdup. A sample of predicted results is shown in Table 1. Thus the resistances can be increased until uniform gas distribution is achieved at the sparger. Comparative study of liquid phase mixing time was done for different sparger resistances to capture dynamics of the flow influenced by sparger. This study will help in identifying the flow behavior for various sparger resistances

Table 1. Dynamic gas distribution on the sparger

U_G, m/s	C_2, (1/m)	Instantaneous snapshots of gas holdup on sparger plate captured at time interval of 1 sec 	Time averaged gas holdup
0.2	4e6	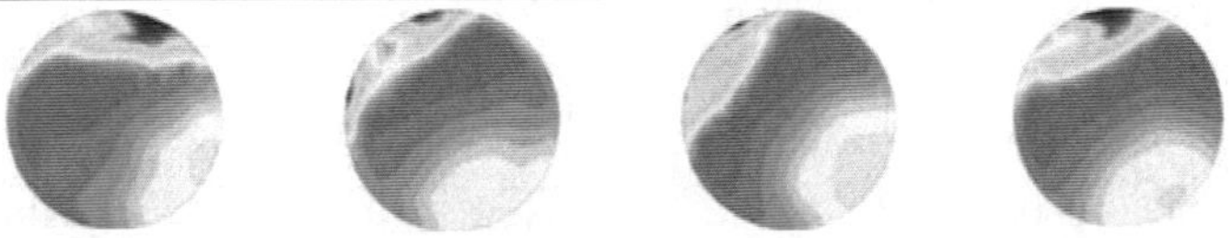	0.318
0.05	4e6	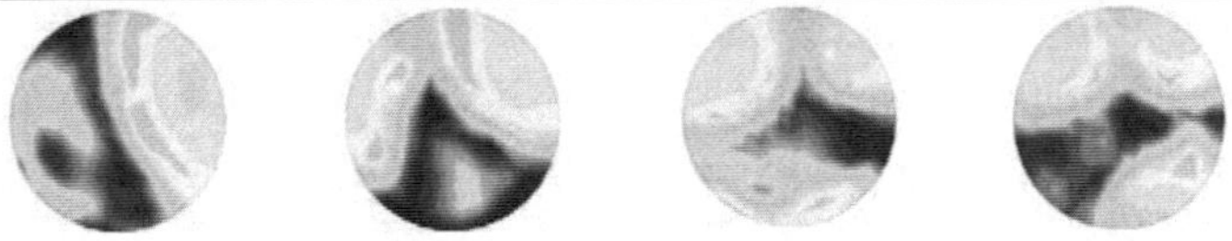	0.103
0.05	4e8		0.112

Conclusion

A CFD simulation for the bubble column reactor considering the plenum section was considered for the first time. Time averaged hydrodynamic properties such as gas holdup, axial liquid/gas velocity showed good agreement with experimental data. The dynamic nature of the flow caused due to pressure gradient at the sparger (porous region) can be studied through liquid phase mixing simulations. The approach, models and results presented in this work will be useful for understanding influence of sparger on gas-liquid flow in bubble columns and will lead to more reliable CFD models and parameters for further work.

References

- Joshi, J. B., (2001). *Chem. Eng. Sci. 55*, 5893.
- Ranade, V. V., (1993). *AIChE symposium series No.293, Vol.89*, 61.
- Rampure, M. R., A. A. Kulkarni and V. V. Ranade, (2007). *Ind. Eng. Chem. Res., doi: 10.1021/ie070079h.*
- Sanyal, J.; V. Sergio, S. Roy, and M. P. Dudukovic, (1999). *Chem. Eng. Sci., 54*, 5071.
- Smith, P. L. and M. Van Winkle, (1958). *AIChEJ, 266.*
- Utikar, R. P., A. A. Sunthakar and V. V. Ranade, (2001). *Indian Chem. Eng. 43*, 16.

324b Predicting Liquid-Solid Flow Regimes in Arbitrarily-Oriented Pipelines with Eulerian Multiphase Simulation

Lanre Oshinowo, Fluent USA, ANSYS, Inc., 16350 Park Ten Place, Suite 204, Houston, TX 77084-5148, Akshat Agarwal, Fluent India Pvt Ltd, Plot No. 34/1, Pune Infotech Park, MIDC, Hinjewadi, Pune, 411057, India, and Sergio Vasquez, Fluent Europe, ANSYS, Inc., Sheffield Business Park, Europa Link, Sheffield, S9 1XU, United Kingdom.

In this paper, we investigate the application of the Eulerian multiphase model to the prediction of liquid-solid transport in horizontal and inclined pipelines. Solid-liquid flow is reported extensively in the literature for horizontal pipes; however, not much has been reported for arbitrarily oriented pipelines. For this work, the solid phase concentration ranges from dilute to dense producing a variety of flow regimes from full solids transport to moving or stationary bed conditions in the pipeline. The turbulent pipeline flow is modeled with the realizable k-e turbulence model with a blended wall function to reproduce the turbulent velocity profile at the wall from the viscous sub-layer through the buffer layer and into the turbulent core. The drag coefficient on the granular solid phase is corrected with the local turbulence dissipation rate. Because particle dispersion in pipeline transport is enhanced by turbulence, the inter-phase turbulence momentum transfer on the dispersion of the particulate phase is included in the analysis and is treated in two ways: a drift velocity correction in the turbulent drag term, and diffusion term in the volume fraction transport equation. The latter approach was found to be comparable to the former in predicting the solids distribution across the pipe and in the moving bed with the additional benefit of eliminating checker-boarding in the solids concentration field due to the weaker pressure gradients relative to the gradient of volume fraction for very dilute slurries. The Eulerian multiphase simulations are conducted in the commercial CFD code FLUENT. The predicted profiles of solids concentration in the pipe cross-section are shown to be in good agreement with correlations for horizontal pipes. From this basis, additional pipe orientations are evaluated and the flow regimes predicted.

324c Simulation of Hydrate Dissociation in Porous Media in Cylindrical Coordinate

Yong Liu[1], Hamid Arastoopour[2], and Matteo Strumendo[2]. (1) Illinois Institute of Technology, 10 W 33rd Street Suite 127, Chicago, IL 60616, (2) department of chemical and environmental engineering, Illinois Institute of Technology, Chicago, IL 60616

This paper describes an axisymmetric model for hydrate dissociation in porous media by depressurization method. A production well of radius of 0.1m is assumed to exist and the gas pressure at the production well is constant during the simulation. A moving dissociation front which separates the total simulation zone into two zones is used. The location of the dissociation moving front expands forward with time. The convective-conductive heat transfer and mass transfer in the gas and hydrate zone together with the energy balance at the moving front are analyzed. The system of governing equations is transformed into new coordinate using a coordinate transformation method. Then the numerical method of lines is used to discretize the governing equations after coordinate transformation. Distributions of temperature and pressure for different well pressure and reservoir temperature are presented. It is shown that the speed of the moving front and the gas production rate are sensitive function of the well pressure and the reservoir temperature.

LES/FDF Model for Turbulent Spray Combustion

Heeseok Koo [1] and Venkatramanan Raman
Department of Aerospace Engineering and Engineering Mechanics,
The University of Texas at Austin, Austin, TX-78712

Abstract

Spray combustion plays an important role in many practical devices including aircraft and internal combustion engines, chemical reactors, and power generation systems. Numerical simulation of spray combustion requires detailed description of gas-phase turbulent mixing and reaction as well as the dispersion of the liquid phase spray droplets. While modeling spray dispersion has received much attention in the past, spray-influenced gas-phase combustion is still modeled using conventional single-phase combustion models. Since the combustion regime in spray combustion is determined locally by the relative rates of spray evaporation and turbulent mixing, different combustion regimes could co-exist in a spray combustor. The objective of this work is to develop a new spray combustion model that will account for these multiple reaction regimes. The filtered-density function approach, originally developed for single-phase combustion, is extended here to spray combustion. A stochastic particle method is coupled with the large-eddy simulation (LES) technique to describe the combustion process. The FDF method, by construction, will allow a direct description of the multiple combustion regimes. This LES/FDF algorithm is used to simulate an experimental spray configuration in order to assess the validity and accuracy of the proposed method. Algorithmic and modeling issues are discussed.

A. Introduction

Spray-based turbulent reacting flows are an important class of engineering systems that have widespread use in many industries including the chemical processing industry. In such flows, a liquid jet is atomized and introduced into a gas-phase carrier. This atomized spray then evaporates and reacts with the carrier-phase. In the aircraft engines and gas turbines, the resulting combustion reactions are highly exothermic and extremely sensitive to the local turbulent mixing process. Typically, combustion systems are classified as premixed or non-premixed flows based on the nature of mixing between fuel and oxidizer at the inlet. Based on this classification, models specific to the combustion regime (i.e., premixed or non-premixed) are developed. Spray systems provide a unique challenge in that both non-premixed and premixed regimes can coexist in the same flow. The nature of flame propagation is determined by, among other factors, the spray dispersion in the gas-phase and the rate of mixing prior to reaction. This leads to a rich variety of flame structures and compounds the modeling process. While much of the work on spray combustion has focused on understanding spray dispersion, a comprehensive model for the combustion process is yet to be formulated.

[1] Corresponding author

To describe the multi-regime spray combustion process, the underlying combustion model should be able to handle all of the combustion regimes. Most mixture-fraction based models will fail in this regard since they describe combustion using a diffusion-flame approach that is not valid in the premixed regime. Similar argument could be used against models for premixed combustion. In this regard, the probability-density function (PDF) approach [1] is uniquely different. Here, the chemical source term is closed by virtue of the PDF methodology allowing different reaction regimes to be described without additional modeling. However, the sub-filter mixing will now have to adjust to the local combustion regime. As a first step, the PDF method is constructed in the context of the large-eddy simulation (LES) methodology. By simulating an experimental spray configuration, the advantages and modeling requirements are then analyzed. The model formulation is described next.

B. LES/FDF/LPT Model Formulation and Numerical Implementation

The filtering approach leads to the modified Navier-Stokes equations that are solved in conservative form using the LES solver. In all known practical LES simulations, the filter is a box filter which implies that the filter is defined by the grid itself. The details of the LES solver used in this work are provided in [2]. Briefly, the sub-filter stress terms are obtained from a dynamic Smagorinsky closure [3]. The LES solver also solves a scalar transport equation using a third-order upwind-biased scheme [2, 4]. This additional scalar which represents fuel-air mixture fraction is used for the feedback mechanism that will be described later.

The Favre-filtered FDF is defined in the context of LES as

$$F_L(\boldsymbol{\psi}; \mathbf{x}, t) = \int_{-\infty}^{+\infty} \rho(\mathbf{y}, t)\, \xi[\boldsymbol{\psi}, \boldsymbol{\phi}(\mathbf{y}, t)]\, G(\mathbf{y} - \mathbf{x})\, d\mathbf{y}, \tag{1}$$

$$\xi[\boldsymbol{\psi}, \boldsymbol{\phi}(\mathbf{y}, t)] = \delta[\boldsymbol{\psi} - \boldsymbol{\phi}(\mathbf{y}, t)], \tag{2}$$

where δ is an N-dimensional delta function for an N-species system and $\boldsymbol{\psi}$ is the random variable in the composition domain. The joint-FDF defined as above yields the following property:

$$\int_{-\infty}^{+\infty} F_L d\boldsymbol{\psi} = \int_{-\infty}^{+\infty} \rho(\mathbf{y}, t)\, G(\mathbf{y} - \mathbf{x})\, d\mathbf{y} = \overline{\rho}. \tag{3}$$

Similarly, the filtered mean of any scalar Q_ϕ can be defined as:

$$\widetilde{Q}_\phi = \int_{-\infty}^{+\infty} Q_\phi(\boldsymbol{\psi}, \mathbf{y}, t) F_L d\boldsymbol{\psi}. \tag{4}$$

Based on this definition, the transport equation for the FDF could be constructed in the case of single-phase flows. When evaporative spray droplets are present, certain additional constraints need to be satisfied. First, the zeroth moment of the FDF transport equation should yield the continuity equation. Second, higher moments should be consistent with the scalar moments equations. For instance, the first and second moment of a scalar can be written as

$$\frac{\partial \overline{\rho}\widetilde{\phi}}{\partial t} + \frac{\partial}{\partial x_j}\overline{\rho}\widetilde{u}_j\widetilde{\phi} = \frac{\partial}{\partial x_j}(D + D_T)\frac{\partial \widetilde{\phi}}{\partial x_j} + \widetilde{S(\phi)} + \overline{\rho}\widetilde{\phi^* W} \tag{5}$$

and

$$\frac{\partial \overline{\rho}\widetilde{\phi^2}}{\partial t} + \frac{\partial}{\partial x_j}\overline{\rho}\widetilde{u}_j\widetilde{\phi^2} = \frac{\partial}{\partial x_j}\left(D + D_T\right)\frac{\partial \widetilde{\phi^2}}{\partial x_j} + \widetilde{\phi S(\phi)} + 2\overline{\rho}\phi^*\widetilde{\phi \dot{\mathcal{W}}} - \overline{\rho}\widetilde{\phi^2 \dot{\mathcal{W}}}, \tag{6}$$

where ϕ^* is the mass-fraction of scalar ϕ present in the evaporated fuel droplet, D is the scalar diffusivity, and D_T is the modeled turbulent diffusivity. To be consistent with these equations, the FDF transport equation should be written as

$$\frac{\partial F_L}{\partial t} + \frac{\partial}{\partial \mathbf{x}}\left(\tilde{\mathbf{u}}F_L\right) + \frac{\partial}{\partial \mathbf{x}}\left(\widetilde{\mathbf{u}'|\psi} F_L\right) = -\frac{\partial}{\partial \psi}\left[\left(M(\psi) + \mathbf{S}(\psi) + \mathcal{G}(\psi)\right)F_L\right] + \dot{\mathcal{W}}F_L, \tag{7}$$

where $\widetilde{\mathbf{u}'|\psi}$ is the sub-filter velocity fluctuation conditioned on the scalar, $M(\psi) = \overline{\nabla \cdot \rho D \nabla \phi|\psi}$ is the conditional micromixing term, and $\mathbf{S}$ is the reaction source term. The conditional velocity term in Eq. 7 is modeled using the gradient-diffusion hypothesis [5]. The term $\mathcal{G}$ accounts for the change in the FDF due to spray evaporation. In order to recover the scalar moment equations (Eqs. 5 and 6), this term should be of the form

$$\mathcal{G}(\psi) = (\phi^* - \psi)\dot{\mathcal{W}}. \tag{8}$$

The FDF transport equation (Eq. 7) is a high-dimensional equation that cannot be solved using finite-volume or finite-difference discretization methods. A stochastic particle method is commonly used in the context of single-phase flows. Here, this algorithm is extended to the spray combustion problem.

In the Lagrangian approach, a large number of notional particles are uniformly distributed in the computational domain. Each of these particles carries a property vector (similar to the spray notional particles) that consists of the particle's thermochemical composition, position, and characteristic weight. The weight of the particles is set initially to be the average weight of the computational cell in which the particles are initialized. If the particles evolve consistently, the weights will provide a measure of the local density at any time-step. These particles then evolve in physical and compositional spaces using stochastic equations.

$$d\mathbf{x}^* = \left[\tilde{\mathbf{u}}^p + \frac{1}{\overline{\rho}}\nabla\overline{\rho}(D^p + D_T^p)\right]dt + \sqrt{2(D^p + D_T^p)}d\mathbf{W}, \tag{9}$$

where $\mathbf{x}^*$ is the position of the particle, the subscript p indicates the corresponding gas-phase value interpolated onto the particle location, $d\mathbf{W}$ is the Wiener diffusion term that mimics turbulent diffusion of the fluid. This term is characterized by a Gaussian process with zero mean and variance of dt, where dt is the time-step used in the computation.

Transport in composition space is due to mixing and chemical reactions. The conditional mixing term ($M(\psi)$) in Eq. 7) needs to be modeled. Several closures have been proposed in the context of single-phase flows [6–11]. Here, IEM model [8] is used due to its computational simplicity. It should be noted that additional constraints may have to be satisfied for turbulent spray combustion. In the present work, this aspect has not been explored. The IEM model is used here only to illustrate the applicability of the LES/FDF approach. The effect of mixing models is being studied currently and will be presented elsewhere.

With this assumption, the particle equations for the composition evolution can be written as

$$d\psi = -\frac{1}{\tau_\phi}\left(\psi - \widetilde{\phi}\right)dt + \mathbf{S}(\psi)dt\,. \tag{10}$$

where τ_ϕ is the scalar mixing time-scale that needs to be specified. A turbulent diffusivity based time-scale is often used [5, 12]. In this work, a dynamic model for the time-scale will be used [13]. It is noted that the choice of mixing model will affect the predictive accuracy of the LES/FDF method. Future studies will aim at understanding the sensitivity of simulation results to the mixing time-scale model used.

In spray combustion problems, the evolution of the liquid spray droplets has to be tracked. A Lagrangian method that evolves the spray number density using a notional droplet ensemble is typically used [14–16]. Here, notional particles with weights corresponding to the mass of the droplets are injected into the computational domain. These particles carry droplet property values and evolve in physical space using a set of ordinary differential equations [14]. The k-th notional particle in the ensemble carries a state-vector consisting of position ($\mathbf{X}_k^p$), velocity ($V_{k,j}^p$), homogeneous temperature (T_k^p), and a size characteristic such as radius (R_k^p). The evolution equations for the notional droplets are then given by

$$\begin{aligned}
\frac{dV_{k,j}^p}{dt} &= \frac{1}{\tau_k}\left(\widetilde{U}_j(\mathbf{X}^p,t) - V_{k,j}^p\right), \qquad \frac{d\mathbf{X}_k^p}{dt} = \mathbf{V}_k^p \\
\frac{dR_k^p}{dt} &= \dot{R}, \qquad \frac{dT_k^p}{dt} = \dot{T}\left(T_k^p, \widetilde{T}, R_k^p, \mathcal{K}\right),
\end{aligned} \tag{11}$$

where τ_k is the particle relaxation time that is usually obtained from an empirical correlation. $\dot{R}$ is the rate of change of radius that is defined through the evaporation rate. $\dot{T}$ is the rate of change of temperature that is a function of the droplet properties, the gas-phase composition and temperature, as well as the physical properties ($\mathcal{K}$) of the droplet. For the particle velocity update, the local gas-phase velocity has to be interpolated from the Eulerian fixed locations. The particle motion also exerts reverse force on the gas-phase evolution. This force appears as a source term in the gas-phase momentum equations. It should be noted that both the forward interpolation and the backward source term evaluation are computationally non-trivial[16].

The LES, FDF, and LPT solvers are implemented using a temporally consistent coupling scheme. The three solvers interact at each time-step through exchange of filtered-fields. The LES solver provides the velocity and turbulence fields that are used to advanced the LPT and FDF notional ensembles. The LPT droplet experiences drag due to the fluid velocity and in turn imposes a drag force on the fluid. This interaction appears as an acceleration term in the droplet equation (Eq. 11) and as a source term in the LES momentum equations. Droplet evaporation leads to an increase in the mass of the gas-phase that needs to be accounted for in the LES continuity equations and the FDF evolution equations. While evaporation appears directly as a source term in the LES continuity equation, incorporation of this effect in the FDF ensemble requires more attention.

Since evaporation causes mass addition and FDF particles represent gas-phase fluid element or some fluid mass, it appears natural to treat evaporation as a source of new particles. In principle, every filter-volume with non-zero evaporation source will lead to addition of particles. However, this approach was found to increase the ensemble size substantially resulting in computational intractability. To avoid this issue, a mass distribution procedure is followed here. The evaporation source is distributed among the existing particles in a given filter volume. This is consistent with the Eulerian-Lagrangian coupling typically used to simulate droplet evaporation. In addition, it obviates the need to add particles to the FDF ensemble. However, this process does introduce diffusion in composition space. The effect of this implementation needs to be tested and will be the studied elsewhere.

C. Simulation of Spray Evaporation Experiment

To test the feasibility of the LES/FDF approach for spray combustion, a series of spray flame configurations were simulated. In this presentation, we include results from the simulation of an evaporating but nonreacting spray system [17]. In this experiment, liquid acetone is injected into carrier air. A portion of the droplets evaporate before exiting the inlet pipe. The droplet-laden premixed jet issues as a round jet with diameter 9.8 mm with a Reynolds number of 15800. Other experimental conditions and parameters listed in table 1.

Jet Reynolds number	15,800
Jet diameter (D)	9.8 mm
Carrier air flow rate $\dot{M}$	135 (g/min)
Liquid fuel injection rate	7.0 (g/min)
Integrated vapor flux at nozzle exit	5.9 (g/min)
Sauter mean diameter of droplet at jet exit	13.7 (μm)
Gas temperature at nozzle exit	275 (K)
Co-flow velocity	3.0 (m/s)

Table 1: Parameters of non-reacting spray jet simulation.

The simulation is carried out in Cartesian coordinate system to minimize discretization errors near the center of the pipe. The domain size in the simulations was set to $30D \times 12D \times 12D$, where D is the diameter of the droplet-laden inlet jet. A $320 \times 192 \times 192$ grid was used for the LES calculations. Grid clustering was used to resolve the steep gradients in the shear layer close to the jet exit. Roughly 200,000 LPT particles and 300 million FDF particles were used to evolve the notional ensembles. Temperature variations in the gas-phase were neglected. The computations were carried out using a MPI-based domain decomposition algorithm on 128 processors. The simulations were run for 3 residence times, where each residence time is measured as the time required by a particle in the co-flow stream to travel the length of the domain. The results shown here were obtained after steady state was reached.

In Fig. 1, the mixture-fraction contours superimposed on droplet locations is shown. The figure shows that substantial gradients in the mixture-fraction exist even the domain exit. In addition, droplets are also found to exit the domain without undergoing complete evaporation. The incoming droplets have considerable inertia and do not follow the fluid path. The Stokes number and inertia decrease as the droplets evaporate. The droplets were found follow the fluid path beyond x/D = 15.

Fig. 2 shows the distribution of FDF particles in each cell. The grid is clustered in such a way that

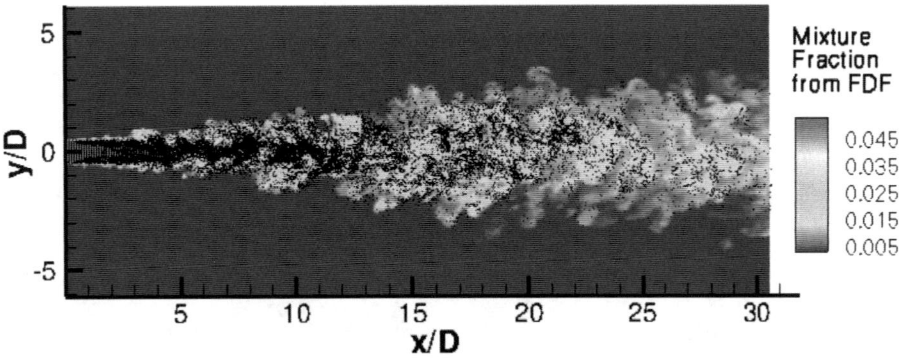

Figure 1: Instantaneous distribution of mixture-fraction superimposed on representative droplet locations.

the cell volumes are smaller near the jet exit. Each FDF particle corresponds to a fraction of the fluid mass in a given filter volume. When the filter volumes are smaller, the number of particles should also be less in order to ensure that the total weight of the FDF particles is roughly equal to the cell mass. It is also possible to maintain a constant particle number in each filter volume. To enforce this, the weight of the particles have to be changed. However, prior experience suggests that such an approach leads to additional spatial errors [13, 18]. For this reason, the FDF particles are assumed to have constant weights. It should be noted that this approach will be ineffective when combustion creates large density changes. In this scenario, many filter volumes will have too few particles leading to large statistical errors. A consistent variable weight algorithm is being formulated currently.

Fig. 3 shows the radial profile of mixture-fraction at different axial locations. Since mixture-fraction does not contain any chemical source terms, the LES-based scalar profile should agree with the FDF-based profile. Minor differences in the profiles can be attributed to the differences in the numerical algorithms used in the Eulerian LES and Lagrangian FDF solvers. The plot also includes profiles of the second moments of mixture-fraction. The RMS (root mean square) of mixture-fraction variance was computed as $\langle \overline{Z}^2 \rangle - \langle \overline{Z} \rangle^2$, where Z is mixture-fraction, the overbar denotes filtering, and the angular brackets denote time-averaging. The FDF-based RMS profile

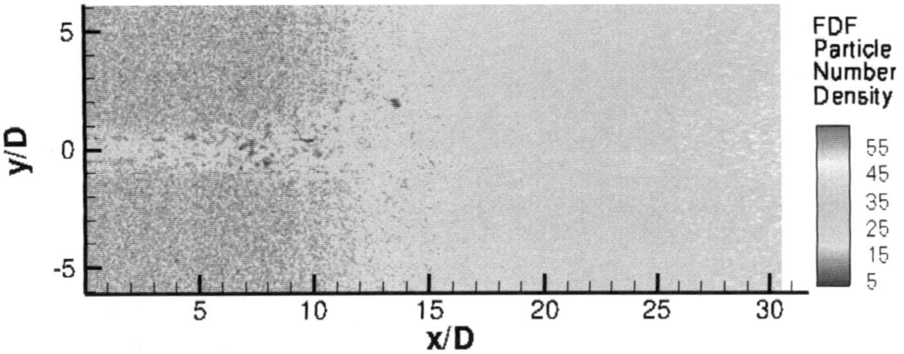

Figure 2: Instantaneous FDF particle number density showing spatial distribution of notional particles.

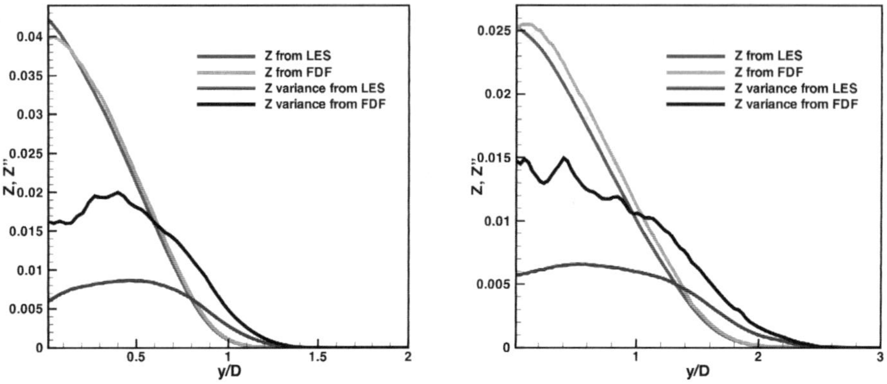

Figure 3: Radial profiles of time-averaged mixture-fraction and mixture-fraction RMS obtained from LES and FDF solvers at x/d=5 (left) and 10 (right).

was calculated as $\langle \overline{Z^2} \rangle - \langle \overline{Z} \rangle^2$. The difference between these two profiles is the time-average of the instantaneous sub-filter variance $\langle \overline{Z''^2} \rangle = \langle \overline{Z^2} \rangle - \langle \overline{Z}^2 \rangle$. The large difference between the two profiles suggests relatively high sub-filter variance and that the sub-filter fluid is highly unmixed in the jet core. Unlike single-phase jets where the mixture-fraction RMS is large only in the shear layers surrounding the jet, the mixture-fraction RMS here is nearly uniform and non-zero even at the jet center. This indicates that significant evaporation occurs throughout the jet cross section.

Fig. 4 shows the centerline velocity profile from the simulation compared with experimental data. The velocity decay is captured with reasonable accuracy near the jet inlet. Further downstream, the simulation velocity decay rate is larger compared to the experimental data. Fig. 4 also shows the radial profile of mixture-fraction compared with experimental data. The jet spread is considerably thinner in the simulations near the jet exit. The agreement is better at x/d=10 indicating that the growth of the shear layer near the jet exit is not being captured accurately. This difference could result from the numerical algorithms as well the sub-filter models used in the LES solver for closing the sub-filter stress terms. These modeling issues are being currently studied.

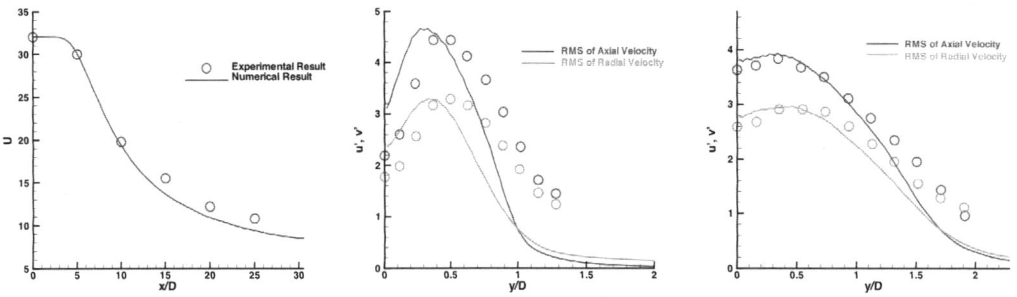

Figure 4: Axial velocity profile along the centerline (left) and radial profiles of velocity at x/d=5 (middle) and 10 (right). The symbols denote experimental data and the lines denote simulation results.

D. Conclusion

A LES/transported FDF method for turbulent spray combustion was formulated. The FDF formulation evolves the joint-composition FDF using the three-dimensional time-dependent LES flow field. The spray population was tracked using a notional droplet ensemble. The evaporation of the droplets and subsequent fuel addition to the gas-phase was accounted for in the FDF model using source terms. The structure of the FDF transport equation is similar to the single-phase formulation but includes additional terms that describe the FDF changes due to evaporation. A particle model was formulated to solve the high-dimensional FDF transport equation. A jet spray experiment problem was used to demonstrate the feasibility of the LES/FDF method. Future work will further verify and validate this new approach by including combustion.

Acknowledgments

The authors would like to thank Prof. Heinz Pitsch and Olivier Desjardin at Stanford University for providing the LES and LPT solvers and for useful suggestions on the spray modeling problem. The authors also thank the American Chemical Society's Petroleum Research Fund for supporting this project through grant PRF# 47067-G9.

References

[1] S. B. Pope, A Monte Carlo method for the PDF equations of turbulent reactive flow, Combustion Science and Technology 25 (1981) 159–174.

[2] C. D. Pierce, Progress-variable approach for large-eddy simulation of turbulence combustion, Ph.D. thesis, Stanford University (2001).

[3] P. Moin, K. Squires, W. Cabot, S. Lee, A dynamic subgrid-scale model for compressible turbulence and scalar transport, Physics of Fluids A 3 (1991) 2746–2757.

[4] B. P. Leonard, A stable and accurate convective modelling procedure based on quadratic upstream interpolation, Computational Methods in Applied Mechanics 19 (1979) 59–98.

[5] P. J. Colucci, F. A. Jaberi, P. Givi, Filtered density function for large eddy simulation of turbulent reacting flows, Physics of Fluids 10 (2) (1998) 499–515.

[6] C. Dopazo, Relaxation of initial probability density function in the turbulent connection of scalar fields, Physics of Fluids 22 (1979) 20.

[7] R. L. Curl, Dispersed phase mixing I. Theory and effects in simple reactors, AIChE Journal 9 (1963) 175–181.

[8] J. Villermaux, Micromixing phenomena in stirred reactors, in: Encyclopedia of Fluid Mechanics, Gulf, Houston, TX, 1986, Ch. 27.

[9] J. Janicka, W. Kolbe, W. Kollmann, Closure of the transport equation for the PDF of turbulent scalar fields, Journal of Non-Equilibrium Thermodynamics 4 (1970) 47–66.

[10] S. Subramaniam, S. B. Pope, Comparison of mixing model performance for non-premixed turbulent reactive flow, Combustion and Flame 117 (1999) 732.

[11] R. O. Fox, Improved Fokker-Planck model for the joint scalar, scalar gradient PDF, Physics of Fluids 6 (1994) 334.

[12] F. A. Jaberi, P. J. Colucci, S. James, P. Givi, S. B. Pope, Filtered mass density function for large-eddy simulation of turbulent reacting flows, Journal of Fluid Mechanics 401 (1999) 85–121.

[13] V. Raman, H. Pitsch, A consistent LES/filtered-density function formulation for the simulation of turbulent flames with detailed chemistry, Proceedings Proceedings of the Combustion Institute 31 (2006) 1711–1719.

[14] J. K. Dukowicz, A particle-fluid numerical model for liquid sprays, Journal of Computational Physics 35 (1980) 229–253.

[15] N. Okongo, J. Bellan, Consistent large-eddy simulation of a temporal mixing layer laden with evaporating drops. part1. direct numerical simulation, formulation and a priori analysis, Journal of Fluid Mechanics 499 (2004) 1–47.

[16] J. Reveillon, L. Vervisch, Analysis of weakly turbulent dilute-spray flames and spray combustion regimes, Journal of Fluid Mechanics 537 (2005) 317–347.

[17] Y. C. Chen, S. H. Starner, A. R. Masri, A detailed experimental investigation of well-defined, turbulent evaporating spray jets of acetone, International Journal of Multiphase Flow 32 (2006) 389–412.

[18] V. Raman, H. Pitsch, R. O. Fox, A consistent hybrid LES-FDF scheme for the simulation of turbulent reactive flows, Combustion and Flame 143 (1-2) (2005) 56–78.

Numerical Simulation of Coal Gasifier for Hydrogen Product and In-situ Fixation of CO_2

LU Jing[a,b], YU Liang[a,b], ZHANG Xiangping[a], ZHANG Suojiang[a*], DAI Wenbin[c]

a. State Key Laboratory of Multiphase Complex Systems, Institute of Process Engineering, Chinese Academy of Sciences , Beijing , 100080

b. China Graduate University of Chinese Academy of Sciences, 100049, Beijing, China

c. Advanced Systems Center, Mitsubishi Materials Corporation, Tokyo, 112-0002, Japan

Abstract

In order to eliminate the global warming effect, it is highly required to develop new technologies with lower or zero emission of CO_2 to the environment. In this work, a fluidized bed coal gasifier with in-situ fixation of CO_2 is simulated by developing a comprehensive two-dimensional unsteady state model. Eddy Dissipation Concept (EDC) model is adopted for describing the turbulent mixing and detailed chemical kinetics simultaneously. The profiles of temperature, velocity and reaction rate, and particle trajectories are predicted and analyzed, also the mechanism of flow and reaction, including the in-situ fixation process of CO_2, is discussed. Comparing with the traditional coal gasifier without CaO, a much higher concentration of H_2 (traditional: 5.78 vol%, present: 65.43 vol%) and lower concentration of CO (traditional: 37.15 vol%, present: 10.51 vol%) and CO_2 (traditional: >45 vol%, present: 14.05 vol%) are obtained.

Keywords: Coal gasification, Numerical simulation, in-situ fixation of CO_2, CaO absorbent

*Corresponding authors. Tel./Fax: +86-10-82627080

E-mail: sjzhang@home.ipe.ac.cn (Suojiang Zhang)

1 Introduction

Coal gasification process is one of the core components of coal utilization technologies, however, vast emission of CO_2 from this process contributes a great to the global greenhouse effect. In order to reduce the CO_2 emission, the usual approach is to capture CO_2 from syngas after coal gasification. However, it suffers from consumption of a large amount of energy and high equipment investment, because the coal gasification and CO_2 fixation are operated in two separated units. Therefore, it is indispensable to develop new efficient technology with in-situ fixation of CO_2.

The in-situ fixation of CO_2 in gasification process was earlier reported in 1990s. In the recent years, continuous coal gasification experiments on bench scale were conducted with CaO as sorbent in the range of 0.1-6.0 MPa and 923 K, the H_2 concentration could reach higher than 80 vol%, only 2 vol% CO_2 and not more than 1 vol% CO were detected.

Although the coal gasification with CO_2 in-situ fixation technology has attracted extensive interests from scientists and engineers, However, the laboratory works of this technology are mainly limited to the exterior studies, such as the process analysis, operating conditions screening, and gas product composition. Calculating Fluid Dynamics (CFD) simulation is an important and effective tool for the development of the novel gasification process. It can supply some flow-chemistry coupling information in the gasifier that is vital for process design and scale-up of units, but is difficult to obtain only through experiments. However, up to now, no work corresponding numerical simulation of the CO_2 fixation-enhanced coal gasification reactor is reported. In this work, a fluidized bed coal gasifier with in-situ fixation of CO_2 is simulated by developing a comprehensive two-dimensional unsteady state model. Eddy Dissipation Concept (EDC) model is adopted for describing the turbulent mixing and detailed chemical kinetics simultaneously. The profiles of temperature, velocity and reaction rate, and particle trajectories are predicted and analyzed, also the mechanism of flow and reaction, including the in-situ fixation process of CO_2, is discussed.

Eddy Dissipation Concept (EDC) model can account for the turbulent mixing and finite-rate chemistry simultaneously by incorporating the detailed chemical reaction kinetics into turbulent reacting flow. Some reactions in gasifer are not appropriate to be looked as infinite chemistry, such as char steam reforming, especially in a steam gasification process. Additionally, in the present process of coal steam gasification with CO_2 in-situ fixation, operating temperature is about 150 K lower than that in the traditional gasifier, which results in the relatively lower reaction rates in the gasifier. Consequently, turbulent mixing and finite-rate reaction need to be considered simultaneously, and the EDC model, though computational costly, offers a relatively practical tool for predicting the turbulent chemistry characteristics in coal gasifier with CO_2 in-situ fixation.

2. Model description and simulation method

Numerical modeling of coal gasificaiton is divided into two types according to multiphase flow model: Euler-Langrange method and Euler-Euler method. In the present fluidized bed coal gasification process, the pulverized coal and CaO particles entrained in the gas have an almost insignificant volumetric fraction and the particle-particle interactions can be neglected, and therefore, an Euler-Langrange method is used here with the gas phase treated as a continuum and solid particles tracked in a Lagrangian procedure.

In the comprehensive model for the coal gasifier with CO_2 in-situ fixation, besides the conservation equations of mass, momentum and energy, the model also includes particle stochastic tracking model accounting for particle turbulent dispersion model, standard κ - ε model for gas turbulence, EDC (Eddy-Dissipation-Concept) model adopted to describe the turbulence-chemistry interaction and P-1 model used as radiation model. The whole flow field is divided into 3564 standard grid mesh, and the governing partial differential equations for all variables were discretized using the finite volume method. Second-order upwind scheme was adopted here for interpolation in the equations discretization. The SIMPLE algorithm is adopted to solve the set of algebraic equations.

The governing equationsof the model are listed in the table 1.

For the gas:

Mass
$$\frac{\partial \rho}{\partial t}+\frac{\partial}{\partial x_j}(\rho v_j)=S \tag{1}$$

Momentum
$$\frac{\partial}{\partial t}(\rho v_i)+\frac{\partial}{\partial x_i}(\rho v_j v_i)=-\frac{\partial p}{\partial x_i}+\frac{\partial}{\partial x_j}\left[\mu_e\left(\frac{\partial v_j}{\partial x_i}+\frac{\partial v_i}{\partial x_j}\right)\right]+\Delta\rho g_i+\rho_p(v_{pi}-v_i)/\tau_{rp}+v_i S \tag{2}$$

Energy
$$\frac{\partial}{\partial t}(\rho c_p T)+\frac{\partial}{\partial x_j}(\rho v_j c_p T)=\frac{\partial}{\partial x_j}(\frac{\mu_e}{\sigma_T}\frac{\partial T}{\partial x_j})+\omega_s Q_s+n_p Q_{gp}+c_p TS \tag{3}$$

Species
$$\frac{\partial}{\partial t}(\rho Y_s)+\frac{\partial}{\partial x_j}(\rho v_j Y_s)=\frac{\partial}{\partial x_j}(\frac{\mu_e}{\sigma_Y}\frac{\partial Y_s}{\partial x_j})-\omega_s+\alpha_s S \tag{4}$$

Gas turbulence:

$$\frac{\partial}{\partial t}(\rho k)+\frac{\partial}{\partial x_i}(\rho k v_i)=\frac{\partial}{\partial x_j}[(\mu+\frac{\mu_t}{\sigma_k})\frac{\partial k}{\partial x_j}]+G_k+G_b-\rho\varepsilon \tag{5}$$

$$\frac{\partial}{\partial t}(\rho \varepsilon)+\frac{\partial}{\partial x_i}(\rho \varepsilon v_i)=\frac{\partial}{\partial x_j}[(\mu+\frac{\mu_t}{\sigma_\varepsilon})\frac{\partial \varepsilon}{\partial x_j}]+C_{1\varepsilon}\frac{\varepsilon}{k}(G_k+C_{3\varepsilon}G_b)-C_{2\varepsilon}\rho\frac{\varepsilon^2}{k} \tag{6}$$

For the particle phase:

Mass

$$\frac{dm_p}{dt}=S_p \tag{7}$$

Momentum

$$\frac{dv_{pi}}{dt}=\frac{v_i-v_{pi}}{\tau_{rp}}+g_i+\left(v_i-v_{pi}\right)\frac{\dot{m}_p}{m_p} \tag{8}$$

Energy

$$\frac{dT_p}{dt_p}=\left[Q_{pg}+\left(\sum_s h_s-h_p\right)\right]\dot{m}_p \Big/ m_p c_{p_p} \tag{9}$$

Particle turbulent dispersion:

$$v_i=\bar{v}_i+v' \tag{10}$$

Heat transfer:

$$\nabla\cdot Q_r=4\sigma_B\left(aT^4+a_pT_p^4\right)-\left(a+a_p\right)G \tag{11}$$

Turbulence-reaction interaction:

$$\omega_s=\frac{\rho\lambda^{\circ 2}}{\tau^{\circ}\left(1-\lambda^{\circ 3}\right)}\left(Y_s{}^{\circ}-Y_s\right) \tag{12}$$

Particle phase reaction

$$R_{s,r}=A_p\eta_r Y_s p\frac{k_r D_{0,r}}{D_{0,r}+k_r} \tag{13}$$

In the present study, the numerical simulation of 2-D coal gasifier for CO_2 in-situ fixation is carried out using CaO as CO_2 absorbent and H_2O/O_2 as reactant gases. The simulation of the present novel coal gasification is performed with the same experimental apparatus and pulverized coal as used in a traditional circulating fluidized bed coal gasification process for the purpose of comparison. A high reactivity Shenmu coal in the literature is selected. Particle size is assumed to be uniform 50μm.

3. Conclusions

A comprehensive model is developed and applied to the simulation of a novel coal gasification reactor with in-situ CO_2 fixation by the addition of CaO. EDC model incorporating the detailed chemistry into the turbulent reactive flow is adopted for modeling turbulence-chemistry. The simulated profile of temperature, velocity, homogeneous reaction rate and generation/consumption rate of gas species from heterogeneous reactions are presented, and the flow-chemistry coupling mechanism in the coal gasifier is analyzed in detail. Based on this analysis, the mechanism of the in-situ CO_2 fixation is depicted.

Using the same gasification apparatus and the coal with the traditional coal gasification process (Fang's work), the simulated novel coal gasification with the addition of CaO can achieve much higher H_2 concentration (traditional: 5.78 vol%, present: 65.43 vol%) and lower CO_2 concentration (traditional: >45 vol%, present: 14.05 vol%). The simulation indicates that the addition of CaO into the gasifier illustrates an obvious effect on CO_2 fixation.

List of Figures

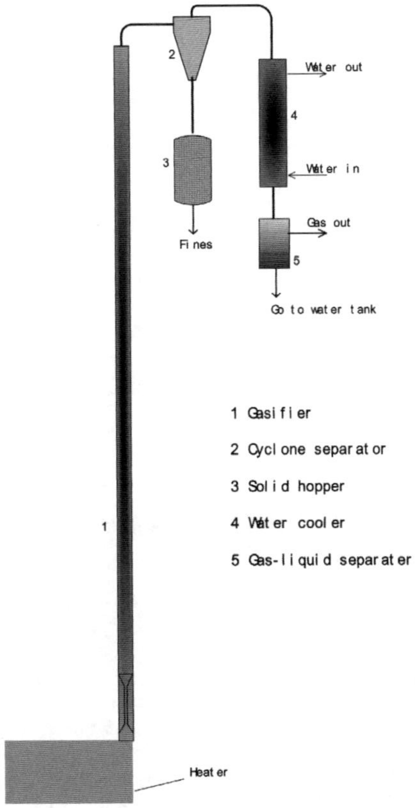

Fig. 1 schematic structure of CFB riser

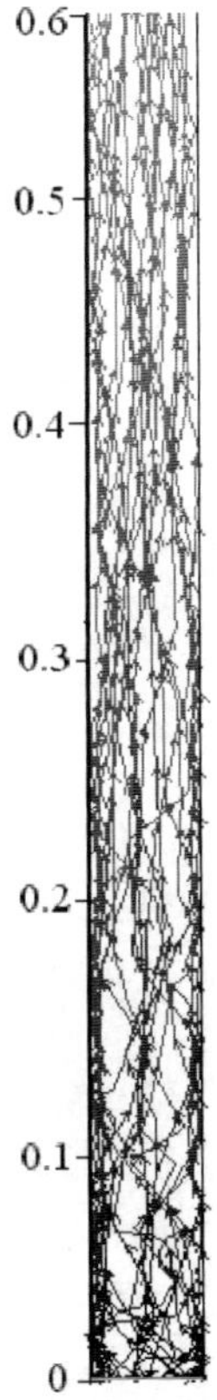

Fig. 3 Particle trajectories up to 0.6 m of the height in the ga:

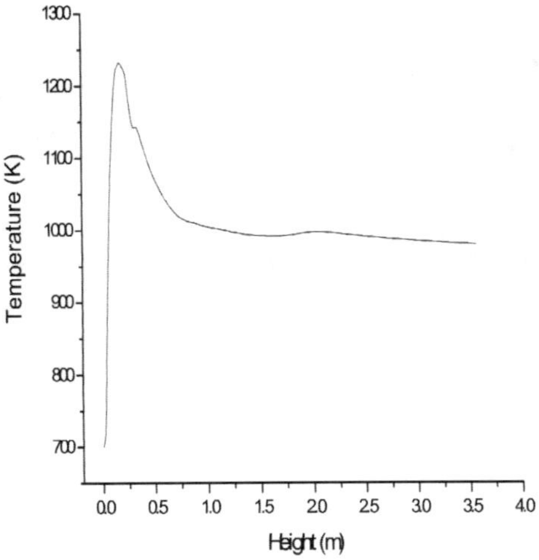

Fig. 4 axial distribution of the gas temperature at center of gasifier (mol/m^3)

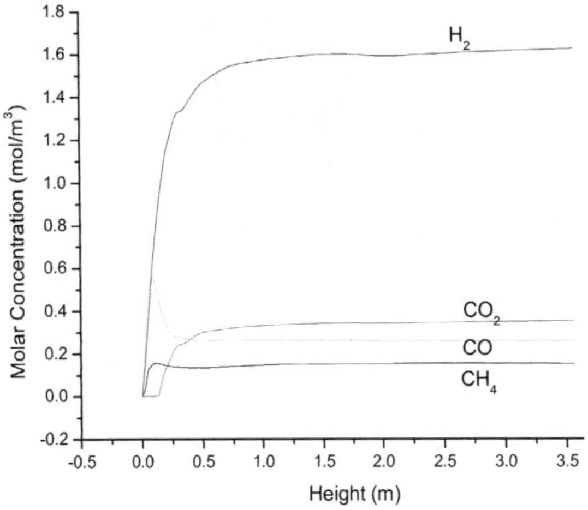

Fig. 5 axial distribution of the gas production molar concentration at center of gasifier (mol/m^3)

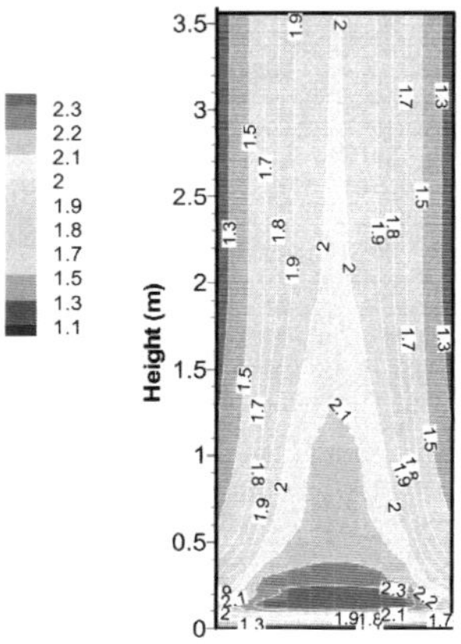

Fig. 2 Profile of velocity in gasifier (m/s)

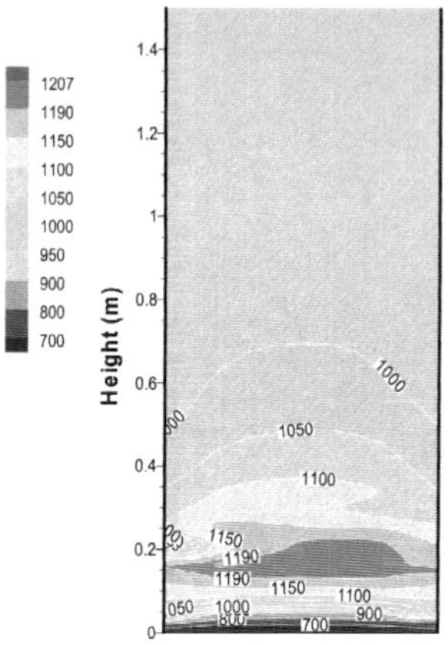

Fig. 6 Temperature profile in gasifier up to the height of 1.5m (K)

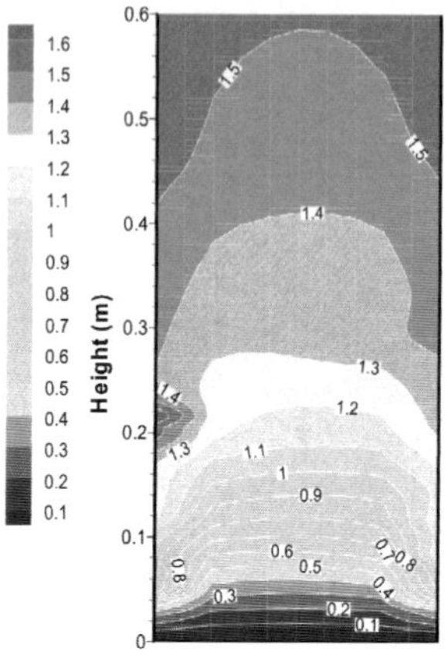

Fig. 7 -a Proflie of H_2 molar concentration in gasifier up to 0.6 m (mol/m^3)

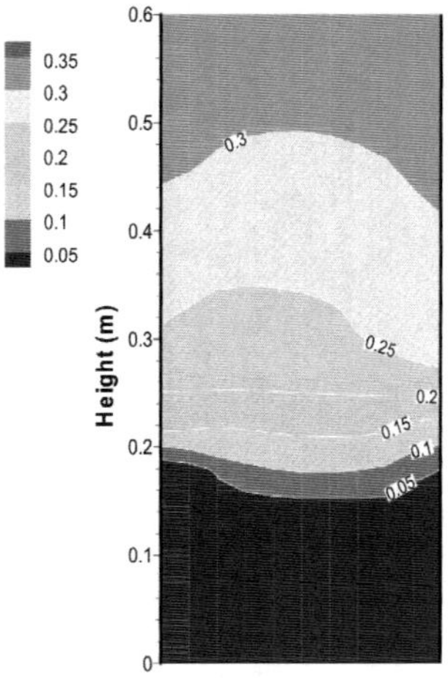

Fig.7-b Proflie of CO_2 molar concentration in gasifier up to 0.6 m (mol/m^3)

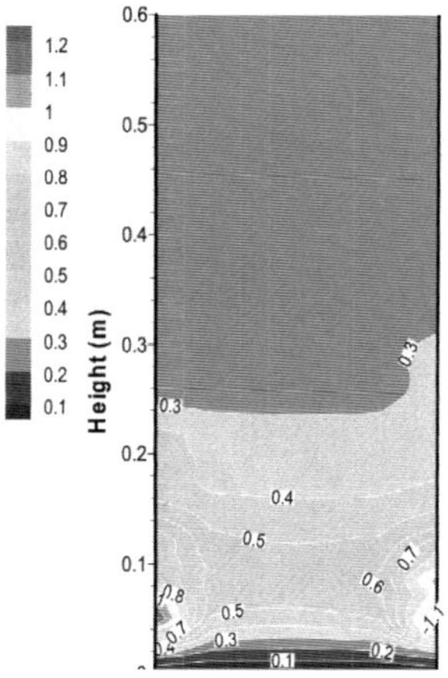

Fig. 7-c Proflie of CO molar concentration in gasifier up to 0.6 m (mol/m^3)

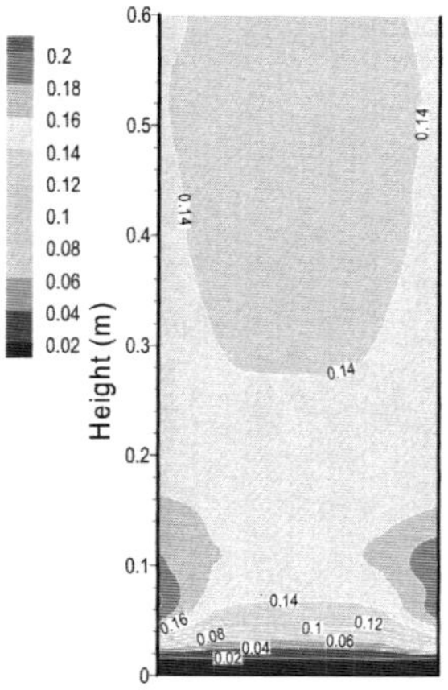

Fig7-d Proflie of CH_4 molar concentration in gasifier up to 0.6 m (mol/m^3)

List of Tables

Table 1 Shenmu coal composition analysis (wt%)

	Shenmu coal	
Proximate analysis (wt%)	Volatile matter	32.15
	Ash	5.99
	Moisture	4.42
	Fixed carbon	the balance
Ultimate analysis (wt%)	C	73.98
	H	4.46
	O	9.70
	S	0.21
Shenmu coal particles	Particle size	50μm

Table 2 Reaction kinetic constants

Chemical reactions	A	E (J/mol)	Unit	β
$CO + H_2O \leftrightarrow CO_2 + H_2$	3.23×10^{7}	4.956×10^{4}	$m^3s^{-1}mol^{-1}$	0
$CH_4 + H_2O \leftrightarrow CO + 3H_2$	3.1005	1.2471×10^{5}	$m^3s^{-1}mol^{-1}$	0
$3C + 2O_2 \rightarrow 2CO + CO_2$	3.51×10^{6}	9.156×10^{4}	$kgm^{-2}s^{-1}Pa^{-1}$	0
$C + H_2O \rightarrow CO + H_2$	0.319	2.0802×10^{5}	$kgm^{-2}s^{-1}Pa^{-1}$	0
$C + CO_2 \rightarrow 2CO$	7.8×10^{-3}	1.3718×10^{5}	$kgm^{-2}s^{-1}Pa^{-1}$	0
$C + 2H_2 \rightarrow CH_4$	6.11×10^{-3}	8.0256×10^{4}	$kgm^{-2}s^{-1}Pa^{-1}$	0
$CaO + CO_2 \rightarrow CaCO_3$	5.771×10^{5}	1596	$kgm^{-2}s^{-1}Pa^{-1}$	0

Table 3 Operating conditions and boundary conditions used in numerical computation

Operating temperature	1000 K
Operating pressure	101325 Pa
Solid feed flow rate (CaO/C)	0.01 kg/s
Mass ratio of coal to CaO in solid feed	1/4
Gas feed temperature	700 K
Gas velocity at entrance	1 m/s
Gas mass fraction at entrance	O_2:steam:N_2 = 0.15:0.45:0.40
Temperature of wall	1100 K

Table 4 Comparison of gas composition

Work	Coal	Temperature	Pressure	Solid feed	Gas feed (wt)	Gas composition (vol%)			
						H_2	CO	CH_4	CO_2
simulated	Shenmu coal	980 K	0.1MPa	Coal/CaO	$N_2/H_2O/O_2$ 0.4/0.45/0.15	65.43	10.51	6.09	14.05
Fang's	Shenmu coal	1123 K	0.1MPa	Coal	CO_2/O_2 0.7/0.3	5.78	37.15	2.13	>45

CFD Application on Hydrodynamic Studies of Multiphase gas-solids Flow in a circulating fluidized bed riser

M. N. Idris
Energy Resource Research Institute and Institute of Particle Science and Engineering
University of Leeds, Leeds, LS2 9JT, United Kingdom
chemni@leeds.ac.uk

Abstract

This paper presents the CFD simulation of hydrodynamic behaviour of gas-solids in a cold pilot plant circulating fluidised bed riser system. It is quite important to have more accurate prediction of flow phenomena for better process performance given the rising cost of feedstock and even decreasing acceptable level of effluent contaminants. The development and use of predictive models for CFB riser reactors are necessary for most small and industrial process scale today. In this work, the relative amount of gas and the solid species as well as the interphase mass, momentum and energy transfer were taken into account. A CFB experimental data was used in these studies and simulation of flow behaviour in the riser using CFD (CFX 11.0) package using k-ε and SST model which are nearly same in values. The flow model is based on an Eulerian-Eulerian description of the phases where the kinetic theory for granular flow forms the basis for the turbulence modelling in the solid phases. In this study, both the gas and velocity profile follows a power law type expression; the gas velocity at the wall is zero. The model predictions are in agreement with the experimental data with FCC catalyst.

Keywords: CFD, Hydrodynamic, Multiphase gas-solids, CFB, FCC riser and Simulation

Correspondence: M. N. Idris, Postgraduate Research Student IPSE-ERRI, SPEME, University of Leeds, LS2 9JT United Kingdom, Tel:+44(0)1133432543

1 Introduction

The gas-solid flow has for many decades been studied quite extensively due to its important applications in nuclear, chemical, petrochemical and petroleum industries. CFB is a core-annulus flow of fluid-solids and solid downflow along the wall of the bed. A CFB feature includes gas and solids distribution and solid acceleration in the bottom zone, the variable solid refluxing at the riser top depending on the exit design, it has a dense-dilute coexistence in the axial voidage profile and the gas and solid flows between the riser and downcomer depending on the total solid inventory, downcomer aeration and the valve design. Circulating fluidized bed (CFB) hydrodynamic models are useful for better understanding of fundamental knowledge about complex multiphase flow gas-solids mixing, scale-up, optimization of process plant system and control. Recently, experimental observations of the flow behaviour have increased the depth of understanding of the mechanisms, which provide the basis for improved field information of the basic concept of conservation equations and constitutive laws for gas-solid flow systems. Conversely, circulating fluidized beds are successfully and widely used in commercial and industrial operations, there remain a need for more research due to the nature and complexity of the gas-solid flow and to establish more fundamental knowledge about the complex hydrodynamic and multiphase flow behaviour. Furthermore, CFB riser operating conditions affect the efficiency of downstream equipment such as cyclones, filters, standpipes, disengager etc.

Hydrodynamic modelling is used for understanding and optimizing plant conditions, but they do not offer an exact level of confidence required to design, a priori, a new commercial plant[1,2].

This paper report an application of advanced computational fluid dynamics (CFD) which has become a viable tool for process simulation of multiphase flow systems. CFD in multiphase flow has become a well accepted and useful tool in modelling of gas-solid flow systems. Application of CFD to validate experimental studies such as this work, will significantly enhance the understanding of the underlying mechanisms, and provide the information to refine conservation equations and the constitutive laws for gas-solids flow systems. Further understanding of the hydrodynamic phenomena, including turbulence characteristics is necessary so that a better closure relationship for turbulence can be developed for CFD modelling using an experimental data[3, 4], while Enwald et al[5] and Hjertager [6] gives a general review of different Eulerian two-phase flow models in fluidization. This process can be significantly accelerated, reducing both time and costs involved. The ANSYS CFX developed model is used to simulate a cold flow laboratory scale circulating fluidized bed.

2. Previous experimental consideration

In this paper two experimental studies were considered. The first one was on solids velocity profile determination. The experiments were conducted with a riser of 15.1m height and 0.10m diameter. The solids used in the riser had a mean particle diameter of 67μm and density 1500kg/m^3 on a superficial gas flow velocity ranges from 3.5-8.0m/s and solid circulation ranges from 50-200kg/(m^{2}s). An ideal gas at ambient conditions was used as the carrier gas in these experiments. The solids flow from the storage tank into the riser distributor region where the solids are fluidised by the auxiliary air. The main air enters through a series of 13mm outer diameter tubes extending 0.30m into the riser bottom to carry the gas-solids suspension up the riser and through a smooth exit into the primary cyclone. The bottom of the cyclone connects to the distributor of the downer. The riser air from the cyclone is sent through a pair of secondary and tertiary cyclones and a bag filter for further cleaning before being exhausted to the atmosphere. The solids from the fast separator are then returned to the switched into the measuring tank for a known time. Reflective-type fibre optic concentration probe were used to determine solids concentration. The active area in the probe tip was approximately 4×10^{-6}m^2, consisting of approximately 8000 emitting and receiving quartz fibres[7], each having a diametre of 15 μm. Details of this probe including its calibration procedure can be found in Zhang et al [8].

The second previous experimental technique considered in this study investigates the flow pattern of different particles size in the riser of a cold-flow circulating fluidized bed. Laser Doppler Anemometer (LDA) techniques were used to experimentally diagnose the flow behaviour of gas-solids systems. LDA were used to measure the mean and fluctuating velocities, and the particle diameter of the dispersed phase. Figure 1 represents the schematic diagram of the CFD boundary domain used in the simulation of the CFB process.

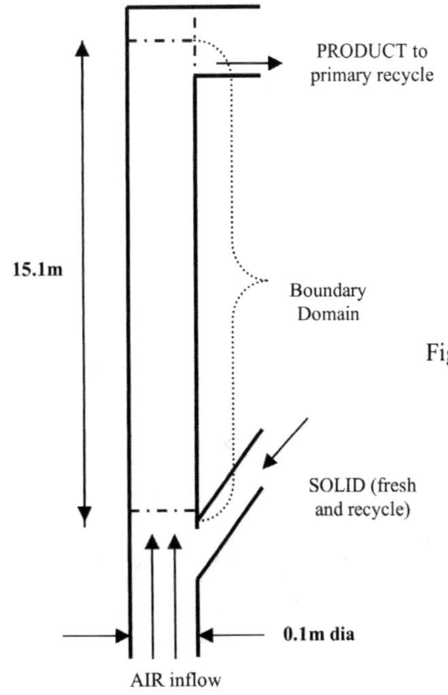

Figure 1: Process flow of CFB system plant

2.1 *Current status of CFD simulation of riser*

The approach for most of the CFD risers simulation are based on steady state[9,10] using the kinetic theory of granular flow as a theoretical model for the solids stress[11].

Usually, these works fixed superficial gas velocity U_g and solid flux G_s without considering their internal relations with solid inventory or imposed pressure drop across the riser, ΔP. However, G_s depend on ΔP and U_g, the former defining the riser hydrodynamics by significantly affecting the height of dense bottom section in risers. CFD application in study of multiphase flow has become a powerful tool in modelling of gas-solids flow systems. Combined experimental and computational studies of the complex multiphase flow behaviour of a circulating fluidized-bed riser system, such as this work, will significantly enhance the understanding of the underlying mechanisms, and provide the information to refine conservation equations and constitutive laws for gas-solids flow systems. The simulation represented in this study was based on the same inlet flow of both gas and solid particles into the riser system.

2.2 *Multiphase gas-solid model*

Two different approaches are used in modelling of gas-solid systems: Lagrangian and Eulerian. The Eulerian approach, which is most commonly used, considers the solids phases as well as the gas as a continuum. The presence of each phase is described by a volume fraction. The presence of each phase is described by a volume fraction. In the Lagrangian approach, the particle control volume is considered to move with the fluid. The dynamics of each particle and its interaction with the fluid are modelled[14]. The Lagrangian approach limits the number of particles, and may not be particularly well suited yet for gas-solid flow systems like fluidized bed. The computational fluid dynamic (CFD) multiphase model for turbulent gas-solids flow is developed in ANSYS CFX are presented in this study. The model uses Eulerian-Eulerian phases, based on the kinetic theory of granular flow as described by[3,4] and an extension of Manger[15]. The Boltzmann equation is used to describe the transport of the distribution functions. The constitutive equations are developed based on the interactions of

the fluctuating and mean motion of the particles[16]. The interactions give rise to effective shear viscosities, which relates the random to the mean motion of the particles[17, 18].
In homogeneous multiphase flow, a common flow is shared by all fluids, as well as other relevant fields such as temperature and turbulence. The transport quantities (with the exception of volume fraction) for that process are the same for all phases.

2.2.1 Eulerian-Eulerian Multiphase Transport Flows

For a given transport process, the homogeneous model assumes that the transport quantities (with the exception of volume fraction) for that process are the same for all phases, i.e.,

$$\phi_\alpha = \phi \qquad 1 \leq \alpha \leq N_p \tag{1}$$

Since transport quantities are shared in homogeneous multiphase flow, it is sufficient to solve for the shared fields using bulk transport equations rather than solving individual phasic transport equations..
The homogeneous model does not need to be applied consistently to all equations, e.g., the velocity field may be modelled as inhomogeneous, but coupled with a homogeneous turbulence model. Alternatively, a homogeneous velocity field may be coupled with inhomogeneous temperature fields[19].

2.3 Continuity equations

The continuity equation for gas phase total mass balance is given by[20]

$$\frac{\partial}{\partial t}(\varepsilon_g \rho_g) + \nabla.\left(\varepsilon_g \rho_g \vec{v}_g\right) = 0 \tag{2}$$

Solid phase total mass balance

$$\frac{\partial}{\partial t}(\varepsilon_s \rho_s) + \nabla.\left(\varepsilon_s \rho_s \vec{v}_s\right) = 0 \tag{3}$$

Where ρ is the density of each phases, ε is the volume fraction, and $\vec{v}$ is the velocity vector

2.3.1 Momentum equations

Momentum conservation gas phase

$$\frac{\partial(r_\alpha \rho_\alpha U^i_\alpha)}{\partial t} + \frac{\partial(r_\alpha \rho_\alpha U^j_\alpha U^i_\alpha)}{\partial x^j} = -r_\alpha \frac{\partial P_\alpha}{\partial x^i} + \frac{\partial(r_\alpha \tau^{ij}_\alpha)}{\partial x^j} + r_\alpha \rho_\alpha g^i + M^i_\alpha \tag{4}$$

Momentum conservation solid phase

$$\frac{\partial}{\partial t}(\varepsilon_s \rho_s) + \nabla.\left(\varepsilon_s \rho_s \vec{v}_s\right) + \nabla.\left(\varepsilon_s \rho_s \vec{v}_s \vec{v}_s\right) = \nabla \vec{T}_s - \beta_B \cdot \left(\vec{v}_s - \vec{v}_g\right) + \varepsilon_s (\rho_s - \rho_g)\vec{g} \tag{5}$$

Where P is the pressure, $\vec{T}_g$, $\vec{T}_s$ are the stress tensors, β is the interface momentum transfer coefficient, and g is the acceleration due to gravity.

Constitutive equations

Definitions

$$\varepsilon_g + \varepsilon_s = 1 \qquad \text{or} \qquad \sum_\alpha r_\alpha + \sum_\beta r_\beta = 1 \tag{6}$$

The ideal gas law is applied to determine the gas pressure where T is the temperature, but this is a cold flow simulation. Then is the gas phase, the pressure of the gas is given as:

$$P_g = \rho_g \overset{\approx}{R} T \tag{7}$$

Assuming a constant density case, the solenoidal total volume flux is given as[20]:

$$\frac{\partial}{\partial x^j}(\sum_\alpha r_\alpha U_\alpha^j) + \frac{\partial}{\partial x^i}\left(\sum_\beta r_\beta U_\beta^i\right) = 0 \tag{8}$$

2.5.3 Interphase momentum transfer

The drag on single particle expressed in terms of dimensionless drag coefficient is given as:

$$D = \frac{1}{2} C_D \rho_f A_p \left|U_f - U_p\right| \left(U_f - U_p\right)$$

Where C_D is the non-dimensional drag coefficient

Assuming that the solid particles are of fixed diameter, and then the volume of single particle is given as:

$$V_P = \frac{\pi d_p^3}{6} \tag{9}$$

Where α= continuous phase, and β = dispersed phase and the surface area of single phase are:

$$S_P = \pi d_p^2 \tag{10}$$

Then the number of particle per unit volume is given as:

$$n_p = \frac{\text{volume occupied by particles in unit cell}}{\text{volume single particle}} \tag{11}$$

$$\Rightarrow n_p = \frac{r_p}{V_p} = \frac{6 r_p}{\pi d_p^3} \tag{12}$$

And

$$\mathrm{Re}_{\alpha\beta} = \frac{\rho_{\alpha\beta} \left|\underline{U}_\beta - \underline{U}_\alpha\right| d_{\alpha\beta}}{\mu_{\alpha\beta}} \tag{13}$$

Where Re is the Reynolds number[20], is the CFX Reynolds number used in the CFX-Solver environment.

The radial velocity of the gas profile can formally be represented by an Ostwald-de Waele[21] type equation:

$$U_g(r) = U_{g,av}\left(\frac{3n+1}{n+1}\right)\left(1 - \left(\frac{r}{R}\right)^{n+1/n}\right) \tag{14}$$

The local solid holdup in a radial region is given as

$$\varepsilon_r = \frac{n_i V_{so}}{\left(\frac{V_{plane}}{10}\right)} = 10\frac{n_i}{\sum_{i=1}^{10} n_i}\varepsilon_a = \frac{V_s}{V} \qquad (15)$$

3 Numerical parameters, initial and boundary conditions

The riser of the CFB is duly considered in this simulation because of its proximity effect in cracking reaction system as applicable in fluid catalytic cracking unit of a refinery. The grid was made in the CFX-Mesh based on unstructured meshing. The grid is uniform in the axial direction, but in the radial direction a non-uniform grid is used in order to have smaller control volumes close to the walls, where the volume fraction and velocity gradient increase. The gas was introduced as a homogeneous phase over the complete bottom region (over an area equivalent to be sum of the cross-sectional area of the riser). A pressure boundary condition was applied to the top of the riser column. A standard no-slip boundary condition was applied at the walls. The time stepping used in the simulations was 100 steps at $5*10^{-5}$s, 100 steps at $5*10^{-4}$s, and 200 steps at $5*10^{-3}$s. This is more significant to reach steady state, which was indicated by a situation in which all of the variables remain constant.

4 Simulation results and comparison with experimental data

The riser section of the CFB was modelled and simulated in a three dimensional Cartesian coordinate system using ANSYS CFX 11.0. The refined unstructured mesh show assumed adequate for these simulation, as shown in Figure 2 and Figure 3. Figure 3 shows the influence of the grid refinement on the velocity profile of solid phase I. The effect of mesh refinement is shown in Figure 4

4.1 Gas velocity profiles

The CFX simulation was performed initially on an ideal gas flow into the riser system based on the entering flowing superficial velocity of 6.4m/s. The fully developed profile was at entrance length of 2.162m of velocity 6.895m/s. Figure 5, 6 and 7 show comparisons of gas velocity profiles of simulated and experimental data. The profiles fit very well with the Ostwald-de Waele correlation; it shows that no downward gas flux can be detected even when solid particle is assumed refluxing at the wall of the riser system.

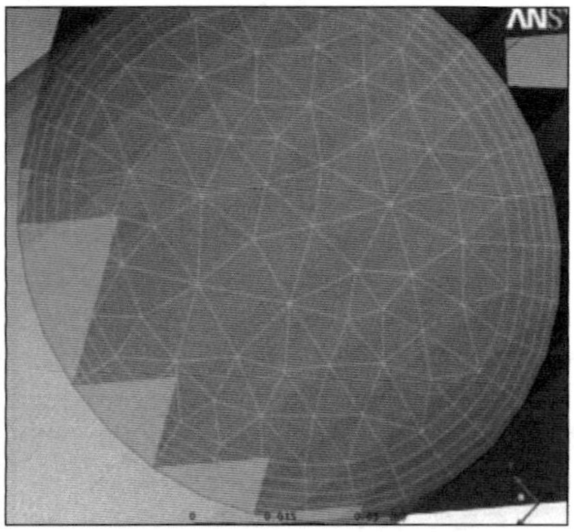

Figure 2: CFX-Mesh 1 (course mesh)

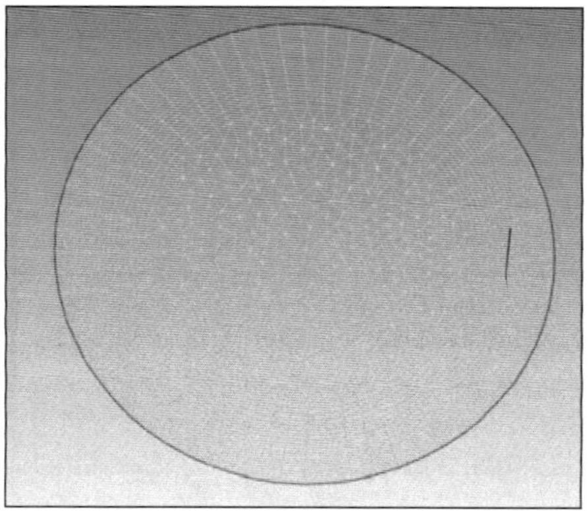

Figure 3: CFX-Mesh 2 (refined mesh) improved used in the simulation

4.2 Particle velocity profiles

Simulation was performed with particles of size 67μm. Measured axial velocity in the free zone of the riser. In this simulation, the boundary conditions for inlet and outlet are based on the schematic diagram outlined in Figure 1. Typical core-annulus flow behaviour for the particulate phases was observed for all flow conditions. The results showed that in the central part of the riser, the particles flow upward with an almost constant velocity. In the wall region, the particle velocity decreases continuously from the core toward the wall. Close to the wall, a low particle velocity and, in some other cases, reversal flows of particles were observed. The data shows that when the superficial gas velocity increases, the relative velocity between the particle groups decreases slightly. This can be attributed to the increase in particle collisions and a more chaotic flow of high gas velocities, the particles show a more

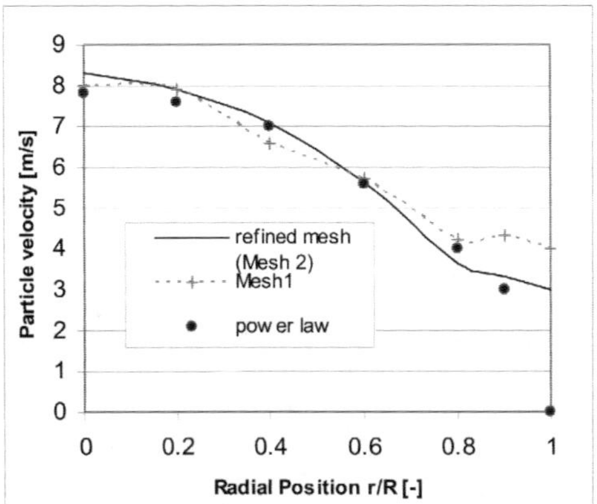

Figure 4: Grid refinement effect on particle

flat solid velocity profile and, in some cases, the maximum velocity deviates slightly from the centre of the riser. The simulated results show some slight deviation from the experimental data, but there is a good comparison in them. These deviations in velocity profile can be said to be the differences in actual and simulated geometry as well as different particle size distribution. This is represented in Figure 8, 9 and 10. In general, the particle velocity is more uniform in the upper section than in the lower section of the riser system and at all radial positions and is higher in the centre than in the wall region of the riser at all axial locations.

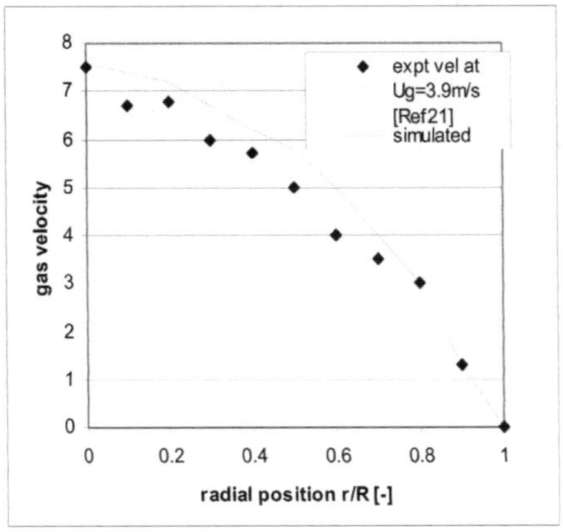

Fig 5 Radial velocity profile at Ug=3.9m/s

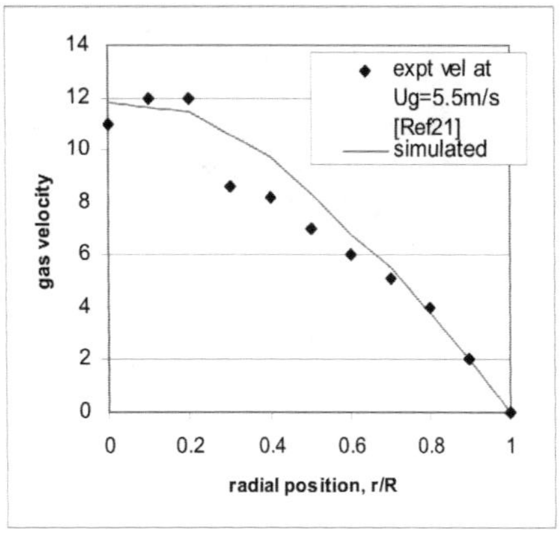

Fig. 6 Radial velocity profile at Ug=5.5m/s

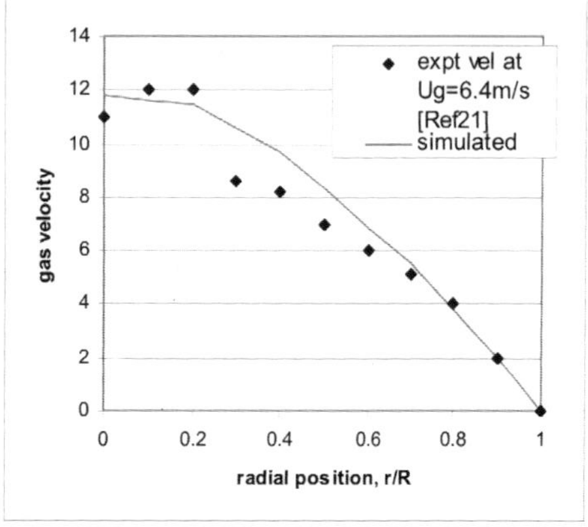

Fig 7 Radial velocity profile at Ug=6.4m/s

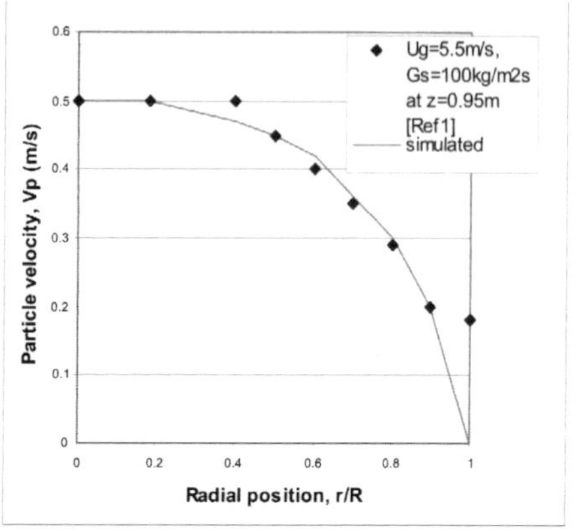

Fig. 8 Particle velocity profiles at z=.0.95m

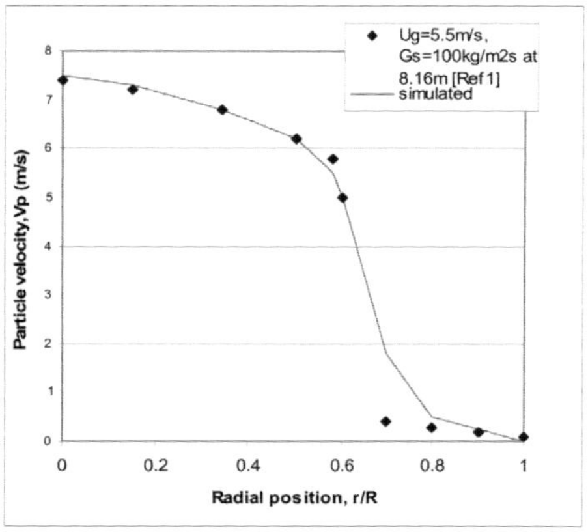

Fig. 9 Particle velocity profiles at z=8.16m

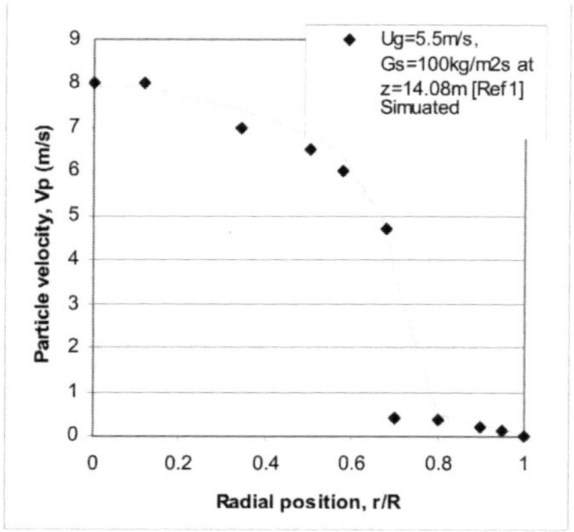

Fig. 10 Particle velocity profiles at z=14.08m

4.3 Solids Holdup in the Fully Developed Region

Under different operating conditions, the apparent solids along the riser obtained from the pressure gradient determined, the overall average solids holdups in the fully developed region, $e_{s,d}$, from the computational simulation was compared with the experimental study as shown in Figure 11. With increase in superficial gas velocity, U_g, there is corresponding decrease in the overall average solids holdups. In flow studies, the superficial gas velocity is greatly an important factor, which has a significant influence on the solids distribution. The simulated values show a good agreement with the experimental data as shown in Figure 12.

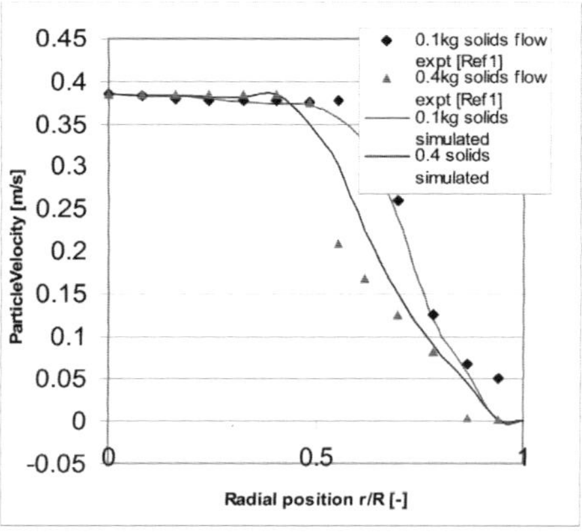

Figure 11: Solids velocity profiles, V_{sup}=0.7m/s

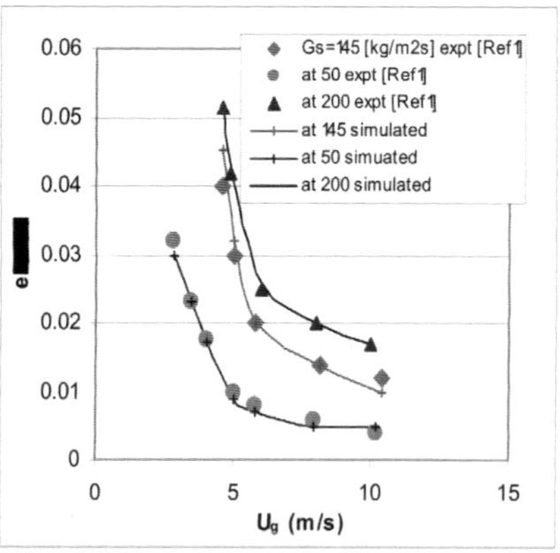

Figure 12: Comparison of experimental and computational value of solids holdup in the fully developed region

5 Conclusions

Computational fluid dynamics package was used to study the multiphase hydrodynamic behaviour of gas-solid flow in the CFB riser system. The model is based on an Eulerian description of the phases where the kinetic theory for granular flow forms the basis for the turbulence modelling in the solid phases. The CFD simulation result was compared with experimental results reported in the literature. The simulation was performed different superficial gas velocities and initial solids concentrations. Although emphasis is given to study the fluctuating behaviour of the gas-solids flow system. We found the simulation to be in good agreement with the experimental data.

The CFD simulation is shown to be able to capture the large scale fluctuation which evidently found to be present in circulating fluidised bed riser system. The deviation observed in axial and radial large scale fluctuations may be due to the experimental data used for the simulation. By adding the large scale-scale fluctuations, a correct order of magnitude is obtained, although some discrepancies were observed. The results show that the axial fluctuations are much larger than the radial fluctuations and hence we can say that in circulating fluidised bed system this behaviour is not isotropic. These provide a realistic description of particle size distribution in gas-solids systems. A computational study of the riser section of the circulating fluidized bed was performed using different velocity flow using FCC particle of size 67μm. The net flux profiles determined show a fairly good agreement with the experimental data. Best agreements were obtained in the upper part of the CFB riser.

6 List of symbols and Abbreviation

$a =$ average value
α or g =gas phase
β or s =solid phase
$\rho =$ density
$\varepsilon_r =$ solid holdup (= solid volume/total volume of the system), %
$U =$ terminal velocity
$V_{so} =$ volume of single solid particle, m^3
$\Gamma =$ Interphase diffusion coefficient
$N_p =$ Number of Phase
$n_i =$ number of solid particles, dimensionless in radial location i =1,2,3,4, ...10
LDA = Laser Doppler Anemometer
S = Source term
SST = Steady State Turbulent Model
$\phi =$ transport quantity

Acknowledgement

The authors would like to express their sincere thank to the Petroleum Technology Development Fund (PTDF), Federal Ministry of Energy, Abuja, Nigeria for the scholarship support in this postgraduate research work.

References

[1]Huang W et al (2007) 'Hydrodynamics and Flow Development in a 15.1m Circulating Fluidised bed Riser', Chem. Eng. Techno: 30, No.4, 460-466

[2]Mathiesen V. Solberg T, and Hjertager B. H (2000) *'An experimental and computational study of multiphase flow behaviour in a circulating fluidized bed'* International Journal of Multiphase flow 26, 387-419

[3]Godfroy Larin, Gregory S. P and Jamal Chaouki (1999) *'Radial Hydrodynamics I Risers'* Ind. Eng. Chem. Res, 38, 81-89

[4]Gidaspow, D., (1994), 'Multiphase Flow and Fluidization' Academic press, Boston

[5]Gidaspow, D., Jung, J., Singh, R. K., (2004), *'Hydrodynamics of fluidization using kinetic theory'* an emerging paradigm 2002 Flour-Daniel lecture. Powder Technology 148, 123-141
[6]Enwald, H., Peirano, E., Almstedt, A. E., (1996) *'Eulerian two-phase flow theory applied to fluidization'* Int. J 36 (4), 523-538
[7]Hjertager, B. H (1997) 'Computational Fluid Dynamics (CFD) analysis of multiphase chemical reactors' Delft University of Technology, March 20-24, Zurich, Switzerland
[8]Zhang H. et al (1998) 'Powder Technology', 100, 260
[9]Dasgupta, S., Jackson, R., Sundaresan, S., (1994), *'Turbulent gas-particle flow in vertical risers'* A.I.Ch. E Journal 40 (2), 215-228
[10]Hrenya, C. M, and Sinclair, J. L., (1997) 'Effects of particle-phase turbulence in gas-solids flows' AIChE Journal 43 (4) 853-869
[11]Hansen, K. G., Solberg, T., Hjertager, B. H., (2004) *'A three-dimensional simulation of gas-particle flow and ozone decomposition in riser of a circulating fluidized bed'* Chemical Engineering Science 59, 5217-5224.
[12]Han and Chung (2001) *'Dynamic modelling and simulation of a fluidized catalytic cracking process'* Part I: process modelling. Chemical Engineering Science 56, 1951-1971
[13]Davidson J. F and Harrison D., (1964) *'Fluidized solids'* Cambridge university press, London
[14]Mathiesen V. Solberg T, and Hjertager B. H (1999) *'Experimental and computational study of multiphase Gas-Particle flow in a circulating fluidized bed riser'* AIChE Journal vol. 45, No. 12, 2503-2518
[15]Manger, E., (1996) 'Modelling and simulation of gas-solid flow in curvilinear coordinates' A PhD thesis, Telemarh Institute of Technology, Norway.
[16]Yang, N., Wang, W., Ge, W., Wang, L., Li. J., (2004) Simulation of heterogeneous structure in a circulating fluidized-bed riser by combining the two-fluid model with EMMS approach. Industrial and Engineering Chemistry Research 43, 5548-5561.
[17]Nieuwland, J. J., van Annaland, M., Kuipers J.A.M., van Swaaji, W.P.M., (1996) *'Hydrodynamic modelling of gas-particle flows in riser reactors'* AIChE., New York, pp. 193-198
[18]Tsuo, Y.P., Gidaspow, D., (1990), *'Computation of flow pattern in circulating fluidized beds'* AIChE, Journal 36 (6), 885-896
[19]Drew J and Passman A, 1993 *'Theory of Multicomponent Fluids'*, Springer Publishing
[20]ANSYS CFX-Solver Theory Guide. ANSYS CFX Release 11.0: 1996-2006 ANSYS Europe, Ltd
[21]Derouin C., et al (1997), 'Hydrodynamics of Riser Units and Their Impact on FCC Operation', Ind. Eng. Chem. Res., 36, 4505-4515

388b Erythrocyte Migration In A Flowing System Bounded By Particle-Free Fluids

Christian P. Aucoin and Edward Leonard. Chemical Engineering, Columbia University, 500 W. 120th St., 801 Mudd Dept of Chemical Engineering, New York, NY 10027

The phenomenon of particle migration in microfluidic flows has long been studied and modelled. However, the application of such learning to practical devices has been limited. Of particular interest is the behaviour of red blood cells, or erythrocytes, in microfluidic devices due to their therapeutic device applications. Recent advances in micro-fabrication techniques has allowed for the application of such learning to a vast array of uses. The primary interest in our laboratory is that of a microfluidic membraneless kidney. Currently, over 400,000 US patients suffer from End Stage Renal Failure Disease, or total kidney failure. The dialysis treatment for these patients requires several hours per week of patient time at a treatment facility, induces a high-amplitude cyclic level of water and toxins in the patient's body, and generates discomfort during treatment due to the required pressure and rate of blood flow. Current treatment entails pumping blood through a membrane filter to remove water and returning the blood to the body over a 3-4 hour treatment 3 times per week. The filters tend to lose permeability over time, and the large surface interface provokes a large set of chemical reactions with blood components. The possibility of treating a patient's blood more steadily and outside of a clinic is truly an intriguing and desirable concept. However, current efforts to extend membrane technology to longer periods of dialysis, outside of the clinic, encounter two difficulties: (1) They require intermittent blood access which is painful, difficult, and potentially dangerous. (2) They do nothing to lessen either membrane fouling with concomitant losses in permeability or problems of biocompatibility occasioned by extensive exposure to artificial surfaces. We have proposed a wearable, indirect dialysis system that is permanently connected to the circulation. It employs direct (membraneless) diffusive contact between blood and a miscible "sheathing" fluid to extract water and toxins from the blood. The cell-free sheath fluid is then dialyzed and ultrafiltered to remove solutes and water and to force recirculation of desirable macromolecules. The dialyzed sheath is recirculated for further contact with the blood stream. The most important factor, by far, in such a system is preventing blood cells from entering the sheath fluid. Past research shows the tendency for such cells to migrate to the center of a flowing stream, and it is believed that blood cells can be retained with proper system design and operation. The nominal design of the current experimental device has a blood channel 100 microns thick flowing between two identical streams of sheath fluid also each 100 microns thick. Equilibrium of the contacted phases is dependent upon the diffusion coefficient of water and toxins in blood and the sheath fluid and is on the order of seconds for small molecules (water, ions, sugars etc.). The transient diffusion equation between static fluids, originally obtained by Loschmidt, yields a solution that closely follows a finite difference representation of the exchange. The sheath system is something that has intrinsic value for us in the application of studies here, but it is also a system that enables the general phenomenon of blood diffusion to be studied in a way that allows separation of the forced and ordinary diffusion components. Past research of other groups, computational fluid dynamic modelling, and experimental analysis in our own lab have shown that red blood cells should remain in the blood stream in high proportion. However a full understanding of the conditions to minimize cell migration from the center stream is unclear. To date, the study of erythrocyte migration has been performed in systems where the blood is bound by impermeable walls. It is well known that erythrocytes cross pathlines under laminar flow conditions due to either the inertia of the fluid, or, in the absence of inertial effects, shear induced migration relying on the deformability of red blood cells. Earlier studies have not considered the dynamics of the system if the blood is bounded not by walls, but by concentric fluids. One may say that this is tantamount to previous studies, but considering the blood as being diluted by the concentric fluids, this statement would be true only if blood enters the system equally distributed throughout the system. In the studies discussed here, blood enters the system via a central stream between two concentric flowing sheath streams, thus testing the theory of how the blood will treat the concentric fluid layers: as an extension of its original continuous phase or as a separate entity. Will the erythrocytes cross streamlines and migrate toward the center as in previous studies despite its seemingly far distance from the walls (thus negating the wall effects), or will the diffusion of blood away from the center stream (driven by the concentration gradient between the central blood-rich stream and peripheral blood-free stream and by particle-particle interactions) be too great to be compensated by the migrational tendencies of blood aforementioned? It is the purpose of this research to address such concepts. The results of such research have engineering applications in regards to artificial kidneys, plasmapheresis, and other biomedical therapies The purpose of this research is to quantify and model the major conditions that affect erythrocyte migration when blood flows in direct contact with 2 miscible blood-free streams. Also, obtaining basic knowledge of erythrocyte behaviour is vital to microfluidic devices designed around blood flows.

388c Studies on Mass Transfer Using Co-Axial Orifice Turbulence Promoters

Sarveswara Rao Sangita[1], Sujatha Vanapalli[2], Rajendra Prasad Padamata[2], and Asha Immanuel Raju Chaduvula[2]. (1) Chemical Engg, Andhra University, Visakhapatnam, AP, India, Andhra University, Visakhapatnam - 530 003, AP, INDIA, visakhapatnam, 530003, India, (2) Chemical Engg Dept, Andhra University, Visakhapatnam, AP, India, Andhra University, Visakhapatnam - 530 003, AP, INDIA, visakhapatnam, 530003, India

Process intensification is a never ending desire for process engineers. It has been carried out for either for efficient operation or to bring down production cost to a minimum. To that effect, process engineers are developing their process all the time. Electrochemical reactor design is not an exception to this norm. To achieve this task number of approaches has been resorted. Use of turbulence promoter is apart of such an effort. The turbulence promoters are of several kinds. Applications of additives to the reaction system, adaptation of mixing devices in the reactor are the principal techniques. Among them use of baffles, stirrers, liquid jets, spargers, use of various geometries immersed in the flow, and the use of two phase, three phase and multiphase flows are commonly adopted. The list is long and the visualization of a technique is the creativity of a process engineer. Among them, use of geometric objects in the flow streams has been receiving the attention of the several workers. Use of orifices in tubular flow systems is not new but studies on axially placed orifices in a series in meager. An orifice generators form friction both in upstream and downstream sides, thus generated friction continuous to shower their effect on wall mass transfer coefficients to longer lengths along the cell. The flow field generated by this type of promoters could not be explained properly by the earlier workers. The enhancements of mass transfer coefficient are not satisfactory at a distance from the orifice under consideration. Because of this reason use of multiple orifices in series transverse to the flow has been used by several workers in their studies. The studies mentioned above enhanced mass transfer to a reasonable extent, on the other hand the transfer coefficients obtained are not uniform along the length of the reactor. Moreover, they showed oscillatory behavior with distance. The generated oscillations are of decaying nature along the path of the flow. Present work is oriented to study the local variation in mass transfer coefficient and to measure the mass transfer coefficients along inner surface of the cell wall. When a series of orifices are placed transverse to the flow, the generated flow fields are complex and of highly rotational. It will enhance transfer coefficient exponentially. In order to harness these benefits, a series of orifice are placed in the tubular cell of circular cross section with varying distances between the consecutive discs. The selected geometry has exhibited good mass transfer efficiency. In the present study orifice diameter Do and orifice to orifice distance 'h' were varied and the experiments are conducted. They yielded 3 to 12 fold enhancements over the reactor without orifices. Mass transfer coefficient increased with decrease in orifice diameter and orifice-orifice distance. An attempt has been made to identify the best combination of Do and h. they yielded 5mm, 30cm, combination is promising. However, overall energy efficiency of the cell can be obtained only with the variation of dynamic variable and velocity together with cost of promoter under consideration. Correlations were also presented for the relative augmentation factor to velocities for each promoter. Generalized correlations were developed in the present study and are presented here. Overall region: Log (g+) = 11.067 (Re)-0.25 (Do/D)-0.5 (h/Do)-0.08 Upstream region:

Log (g+) = 21.469 (Re)-0.32 (Do/D)-0.41 (h/Do)-0.02 Downstream region: Log (g+) = 3.682 (Re)-0.21 (Do/D)-0.58(h/Do)-0.32 Thus presented correlations and information generated are highly useful for the design and development of energy efficient electrochemical cells.

Investigation of electrostatic charging phenomenon in multiphase flows using the ECVT system

Q. Marashdeh* [1,], Wang Fei [1], A.-H. A. Park [2], and L.-S. Fan [1]
(1) Department of Chemical and Biomolecular Engineering
The Ohio State University, Columbus Ohio, 43210
(2) Department of Earth and Environmental Engineering
Columbia University, New York, NY 10027

Abstract

The applicability of different electrical tomography sensors depends on the interrogating signal as well as the sensor itself. In all electrical tomography sensors, the electrical field distribution plays a major role on the measured boundary electrical property. The electrical field distribution is based on the electrical property distribution inside the imaging domain and the generated electrostatic charges during the process operation. In this study, the electrical capacitance tomography sensor is used to study the effect of electrostatic charges on image reconstruction.

Keywords: Electrostatic charges, Electrical capacitance sensor; Multiphase flow;

Introduction

Among tomography techniques for process control and visualization, electrical based tomography systems are the most prominent due to their low construction cost, high speed, and safety [1]. The development of electrical tomography has been mainly focused on permittivity, conductivity, or permeability imaging. Successful implementation of tomography systems depends on the sensor configuration, the data acquisition hardware, and the algorithm used for reconstruction [2]. In addition, electrical tomography results can be challenged by electrostatic charges present in the imaging domain. Electrostatic charges are common in multi-phase flows and they affect both process control and measurements.

Electrodynamic sensors are generally used for measuring the level of electrostatic charges in a flow system. Measurements obtained from electrodynamic sensors are based on detection of varying electric fields from moving particles. Application of this technique has been focused on measurements of flow rates in multiphase flow systems. A modification of the system has been recently introduced for measuring concentration profiles of dry powders. In this case, the measured signal represents the accumulated charges on sensor plates resulting from static charges in the imaging domain. Although electrodynamic sensors have been applied for flow rate measurements, this technique is not suitable to fully explore the effect of the electrostatic charges in various tomography applications. Thus, an electrostatic tomography (EST) technique is developed to provide 3D real time information regarding the charging phenomenon in multiphase flow systems.

Similar to other electrical tomography systems, an EST system is composed of three main parts: (1) sensor, (2) data acquisition hardware, and (3) reconstruction algorithm. In an EST system, a set of sensors is mounted around the boundary of the process vessel. The sensor plates are designed to measure voltages due to static charges introduced through process dynamics. The acquisition hardware is designed to measure the voltage difference (resulting from static charges in measurement domain) between each pair of sensors.

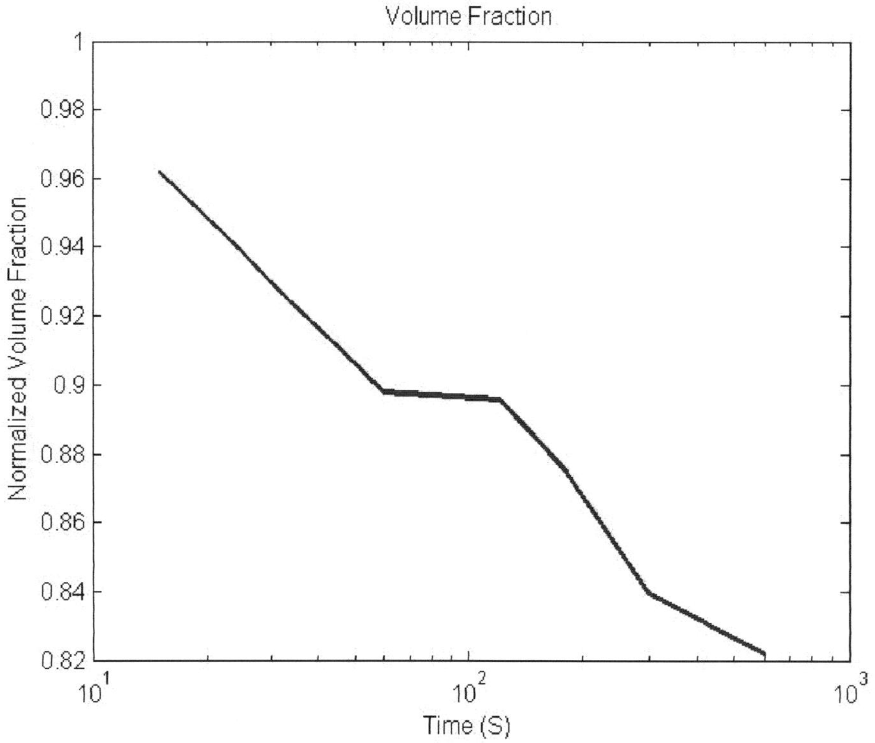

Figure 1 Normalized fraction volume for different times of poppy seed bed fluidization

In this work, the electrostatic charging phenomenon of poppy seeds in a fluidized bed system is investigated with emphasis on the investigation of the effect of static charges on the capacitance measurements of the electrical capacitance volume tomography (ECVT) system.

Experiments

The effect of static charge build up on an ECVT system has been investigated based on poppy seed fluidized bed in dry conditions. The ECVT used is comprised of 12-electrode rectangular sensor arranged in triple planes as shown in [3]. Its performance for 3D volume imaging is examined. The choice of the electrode number is based on the data acquisition system available for experiments, which is, but not limited to, 12 channels. For the rectangular sensor, the electrodes can be arranged, for example in several planes shifted in alignment, in order to distribute the electrical field intensity uniformly in the axial direction and to increase the radial resolution. The radial resolution of the rectangular sensor with this electrode arrangement, thus, equals to 8-electrode sensors per plane. The length of the sensing domain is 10 cm. The image reconstructed on 20x20x20 resolution based on the algorithm described above with volume images digitized in three dimensions. There are 66 combinations of independent capacitance measurements between electrode pairs from 12 electrodes. The results are obtained from a 12-channel data acquisition system (DAM200-TP-G, PTL Company, UK), the image capture speed is of 80 frames per second. The reconstruction process and data post-processing are obtained from a Pentium 4 machine, 3 GHz and a memory of 2 GB. The column size is 0.1 m ID by 1.0 m height.

The ECVT system has been used to image the bed at rest in the initial state. The bed was then fluidized for different durations of time and the ECVT system was again used to measure capacitances of the packed bed similar to the initial state. The difference between readings for different duration of bed fluidization was referred to different levels of electrostatic charge buildup.

Results

In figure 1, the normalized volume fraction of the packed bed is depicted as a function of fluidization time. It is clear from the figure that the normalized volume fraction value decreases as a function of time. In figure 2, the volume distribution of the poppy seeds in the packed bed is depicted. The packed bed imaged for the duration of fluidization is similar; however, the difference in packed bed distribution is mainly due to static charge generation during fluidization. Note the density of the packed bed appears to be less as a function of static charge levels.

Conclusions

In this work, the ECT sensor has been used to study the effect of electrostatic charges generated in a fluidized bed using electrical capacitance volume tomography. The level of static charge generation depending on time and fluidized particles in different phases in the imaging domain changes the value of electric field. Results provided in this work offer volume visualization of the charge distribution in the bed at rest condition. The acquired charge distribution can be used to correct for capacitance measurements.

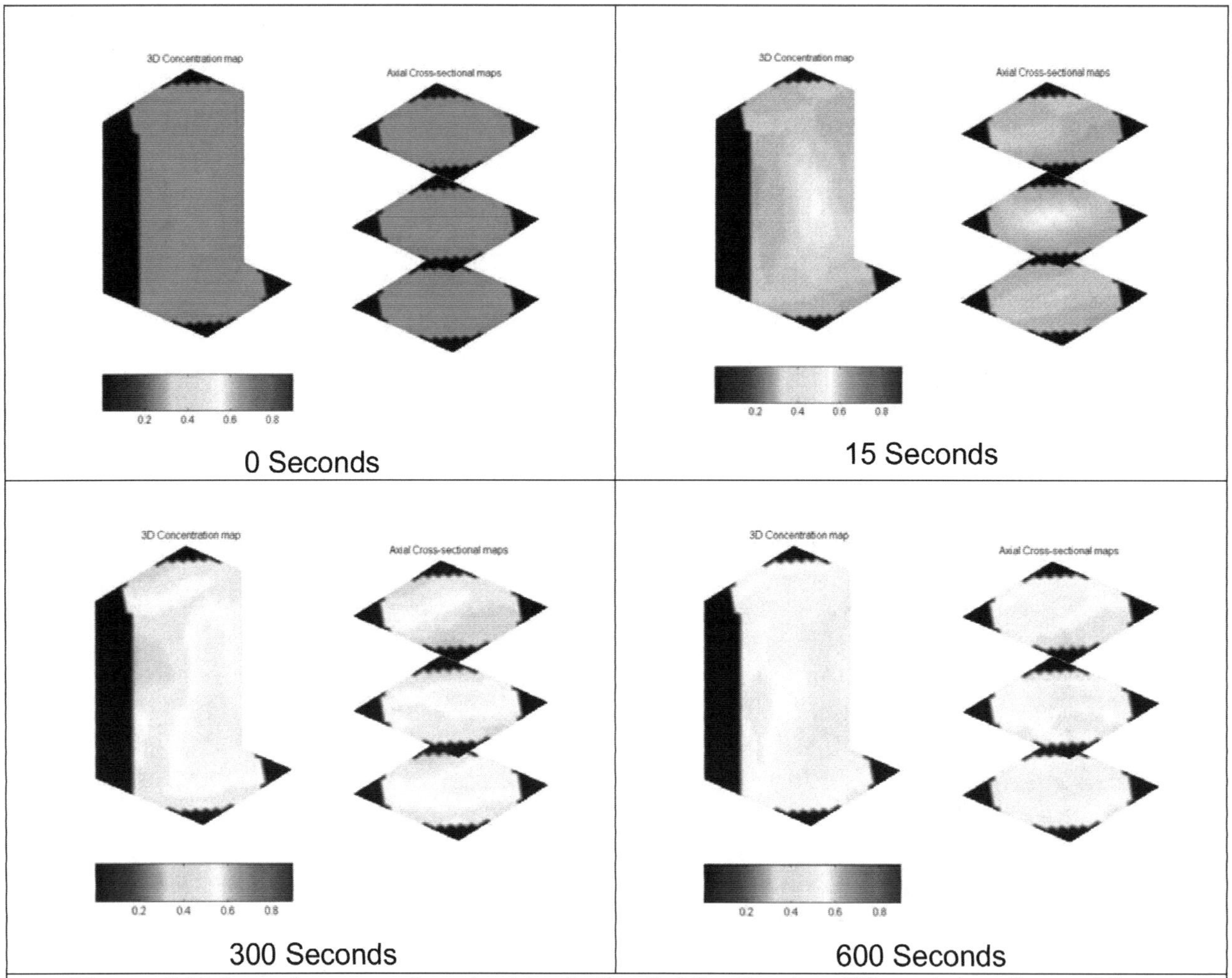

Figure 2 Volume images of packed bed after fluidization for different durations of time

References

[1] R. A. Williams, and M. S. Beck, *Process Tomography—Principles, Techniques and Applications*, (Oxford: Butterworth-Heinemann), 1995.

[2] W. Q. Yang and L. Peng, "Image reconstruction algorithms for electrical capacitance tomography", *Meas. Sci.Technol* , vol. 12, pp. R1--R13, 2003.

[3] W. Warsito, Q. Marashdeh, and L.-S. Fan, "Electrical Capacitance Volume Tomography", *IEEE Sensors Journal* , vol. 7, pp. 525--535, 2007.

388e Augmentation Of Apparent Diffusion Coefficient Of Bovine Albumin And Urea By Increases In Red Blood Cell Concentration During Laminar Flow

Edgar E. Nanne, Christian P. Aucoin, and Edward Leonard. Chemical Engineering, Columbia University, 500 W. 120th St., 801 Mudd Dept of Chemical Engineering, New York, NY 10027

Diffusion coefficients of species in complex fluids are dependent on the dynamic condition under which the diffusion will occur. It has been observed that the apparent diffusion coefficients of bovine serum albumin and urea in a laminar flow of blood are a function of both red blood cell concentration and shear rate. When subjected to shear, red blood cells rotate, carrying fluid next to their surfaces with them. Thus, transport of other species is affected by the cells' rotation. Diffusion coefficients are measured in a microfluidic system in which three congruent streams flow under laminar conditions. The central stream contains the molecular species whose diffusion is being studied and cells at various concentrations (hematocrits). The side stream may or may not contain cells and does not contain the molecule of interest.. Cell-free side streams emulate a blood washing scheme under study, while cell-containing side streams enable the study of solute diffusion within uniform cell concentration and shear fields. The microfluidic flow channel is formed in a block of stainless steel and has an overall channel height of 300 μ m and an aspect ratio of approximately 15. Syringe pumps are used to meter the flows while final concentrations are measured in the exit streams with spectrophotometric techniques. The novelties of incorporating microfluidic techniques include: 1) The opportunity to simulate a wide range of shear flows; 2) the ability to introduce sharp gradients in solute concentrations in middle parts of a flowing stream; 3) do rapid measurments consequent to the short diffusion times experienced in the small distances that characterize the microfluidic channel. Given that the ultimate purpose of this research group is to develop a new artificial kidney that will replace current end stage renal disease (ESRD) treatments, the impact of hematocrit in the transport of different species becomes an important parameter in the final design.

ANALYSIS OF FALLING PARTICLE IN SOLID PARTICLE SOLAR RECEIVER

B.R.Vijayarangan, University of Nevada Las Vegas, Nevada
Samir F. Moujaes, University of Nevada Las Vegas, Nevada

Abstract

An improved correlation for the estimation of drag coefficient in the flow of solid particle in the solar receiver is being presented. Drag coefficient on free falling particles is of considerable importance. In this work, typical evaluations of widely used correlation have been performed. The analysis of these correlations are based on sphericity, void fraction, particle diameter and Reynolds number ranging from 0.6 to 0.98, 0.25 to 0.95, 400 to 650 μm and $10^{-2}<Re<10^{3}$, respectively. Experimental results from four different studies in the literature have been used for comparison with the predictions. A correlation based on Wen and Yu is proposed to predict the entire range of the data sets very well within the average relative error of 19.29 %. The proposed correlation which incorporates the influencing parameters of both void fraction and sphericity can now fill the current void in the literature.

Introduction

Over the past eight decades or so, considerable importance has been given to the development of drag coefficient for particles of spherical and non-spherical shapes. Most of these correlations are expressed to relate the drag coefficient with Reynolds number, sphericity ψ and with some other physical parameters. The correlations/models from the literature are broadly classified into two categories. One group of correlations / models [1-7] are based on a single particle falling/moving in a swarm of particles. The other group of drag correlation/ models [8-10] is based on multiparticle/cluster of particles, giving importance to the effect of porosity or voidage and particle to particle interaction. The behavior of such widely used correlations is analyzed for the present experimental conditions in a solid particle solar receiver which uses particles made from sintered alumina as the heat transfer medium to other process.

The solid particle solar receiver basically works on the principle of receiving concentrated solar energy of the order of magnitude of 2.4 MW/m^2. Figures 1 and 2 show the schematic diagram of the test facility. The design and development of a solid particle receiver is currently in progress at Sandia National Laboratories, Albuquerque, NM, USA. A Solid particle solar receiver is employed to receive concentrated solar energy from a field of heliostats to heat ceramic particles. Thus the heated ceramic particle can serve as a heat transfer and storage medium or as a substrate on which chemical reaction may be performed directly. The solid particles are enclosed in a cavity to absorb concentrated solar radiation and can provide efficient absorption of concentrated sunlight [11-21]. Studies [22-26] show that a large-scale production of hydrogen using solar energy at comparatively lower cost is feasible with a solid particle solar energy receiver concept.

The characterization of the particle flow appears as a first and foremost important step. Drag force models for these particles are often required for wide ranging applications in chemical, mineral and process industries. Inspite of the existence of numerous models/correlations available in the literature for predicting the drag coefficient of spherical/non-spherical particles and are based on sphericity or on volumetric shape factor [1-10], superficial velocities of phases and on other shape parameters. A correlation that can accurately predict the drag coefficient in freely falling curtain of solid particles has not yet been found. To this extent, an evaluation of existing methods is performed in contrast with the free fall of a curtain of ceramic solid particles.

Theory

The forces acting on a single falling particle are classified into three. They are gravitational force, buoyancy force and fluid drag force. Since the gravitational force is very well defined based on Newton's second law, the other two forces, i.e. buoyancy and fluid drag are to be evaluated as accurately as the gravitational force. When particle velocity reaches a constant value, then the aforementioned three forces must be in dynamic equilibrium i.e. its weight must be equal to the fluid drag forces and buoyancy forces together

$$\frac{\pi}{6}d_p^3(\rho_p-\rho_f)g=\frac{1}{2}\rho_f u_{rs}^2\frac{\pi}{4}d_p^2 C_D \qquad \text{..........} 1$$

Where u_{rs} is the relative velocity between particle and fluid. The effective drag can be calculated from Eq. 1

$$C_D=\frac{4d_p(\rho_p-\rho_f)g}{3\rho_f u_{rs}^2} \qquad \text{..........} 2$$

In this work an analysis of widely used models/correlation for the estimation of drag coefficient in solid particle solar receiver is presented.

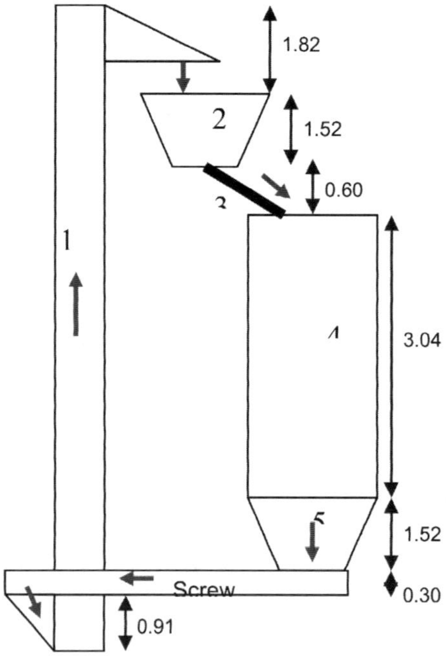

Figure 1. Schematic diagram of experimental facility

1 Bucket Lift to transfer 2000 kg/hr of solid particle to receiver hopper to the lift height between 8 and 10 m
2 Receiver Hopper to hold 0.7 m^3 of solid particles with Load cells
3 Vibration feeder
4 Test Section
5 Catch Hopper to catch solid particles from the test section

Analysis

The solid particles are made to flow within the receiver in the form of curtain, whose dimensions are close to the dimensions of the receiver and slot opening as shown in figure 2. For example, if the slot

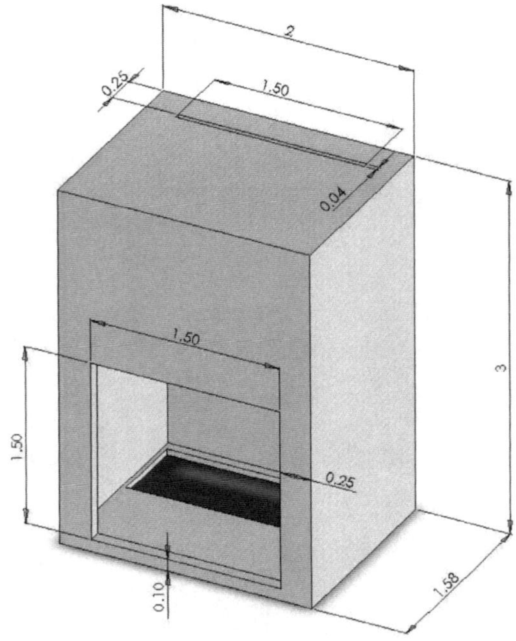

Figure 2. Test section details [32]

opening is 0.04m width and 1.5 m long, then the thickness and width of the curtain takes almost the same dimensions as of slot opening and the height of the curtain is equal to the height of the receiver height as shown in figure 2. To this extent, the curtain height is divided into some equal number of parts, which facilitates, for a better estimation of drag coefficient, on those points. Thus the velocity of an individual particle traversing the curtain from top to bottom can be obtained from the vertical height drop and time taken by the particle to reach each segment. The falling time of a single particle and its respective velocity is accomplished by incorporating selective drag models from the literature and is computed by using a developed code in MATALB to solve a first order ODE which is shown in Eq.3.

$$m_p \frac{du_p}{dt} = (m_p - \rho_f V)g - \frac{1}{2}\rho C_D A u_p^2 \qquad \text{----------------3}$$

The Reynolds number is one of the important non-dimensionless numbers in drag coefficient estimation and is computed based on the particle diameter and relative velocity (u_p- u_f) between the two-phases. Since, the drag coefficient of a particle is strongly influenced by its size, shape and orientation in addition to the viscosity and the densities of the particle and the fluid medium, the selection of the following correlations are based on those parameters.

Widely used existing predictive methods

Haider and Levenspiel [1]

The model by Haider and Levenspiel were the first to develop of universally an applicable and easy to use drag expression for spherical and non-spherical particles in incompressible media. They proposed the following general form of the drag expression for spherical and non-spherical particles.

$$C_D = \frac{24}{\mathrm{Re}}\left[1 + \{8.1716 \exp(-4.0655\psi)\} \times \mathrm{Re}^{(0.0964+0.5565\psi)}\right] + \frac{73.69\,\mathrm{Re}\exp(-5.0748\psi)}{\mathrm{Re} + 5.378\exp(6.2122\psi)} \quad \text{..... 4}$$

The applicability of the correlation is limited to Reynolds numbers range of less than 2.6×10^5.

Ganser [2]

He introduced two shape factors K_1 and K_2 applicable in the Stokes and Newton's regimes, respectively and put forward the following drag correlation

$$\frac{C_D}{K_2} = \frac{24}{\mathrm{Re}K_1K_2}\left[1 + 0.1118(\mathrm{Re}K_1K_2)\right]^{0.6567} + \frac{0.4305}{1+\dfrac{3305}{\mathrm{Re}K_1K_2}} \quad \text{---------5}$$

Where C_D and Re are based on the equal volume sphere diameter, and K_1 and K_2 were found to be unique functions of the sphericity, ψ as shown in Table 1.

Table 1. Expressions for K_1 and K_2 in Eq. 4 in terms of ψ

	K_1	K_2
Isometric shapes	$\left[\left(\frac{1}{3}\right)+\left(\frac{2}{3}\right)\psi^{-0.5}\right]^{-1}$	$10^{1.8148(-\log\psi)^{0.5743}}$
Non-isometric shapes	$\left[\left(\frac{d_n}{3d}\right)+\left(\frac{2}{3}\right)\psi^{-0.5}\right]^{-1}$	$10^{1.8148(-\log\psi)^{0.5743}}$

Swamee and Ojha [6]

Swamee and Ojha developed a correlation based on the literature data of Schulz et al. [31]. Swamee and Ojha employed the equal volume sphere diameter and used the so called Cory shape factor, β, defined as:

$$\beta = c/\,(ab)^{1/2} \quad \text{--------6}$$

Where a>b>c are the lengths of the three principle axis of the particle. Swamee and Ojha's drag expression is as follows

$$C_D = \left[\frac{48.5}{(1+4.5\beta^{0.35})^{0.8}\,\mathrm{Re}^{0.64}} + \left\{\left(\frac{\mathrm{Re}}{\mathrm{Re}+100+100\beta}\right)^{0.32}\frac{1}{\beta^{18}+1.05\beta^{0.8}}\right\}\right] \quad \text{---------7}$$

Eq.6 is applicable in the range 0.3< β <1 and 1<Re<10,000.

Chien [3]

Chien performed a thorough analysis of the data available in the petroleum engineering and processing literature and proposed the following simple expression for drag:

$$C_D = \left(\frac{30}{\text{Re}}\right) + 67.289\exp(-5.03\psi) \qquad \text{--------8}$$

Where, an equal volume sphere diameter is used in both Re and C_D. Though Chien [3] did not present much detail about the fluids and the particles used in the experimental studies, Eq.8 is stated to be applicable in the ranges $0.2 \le \psi \le 1$ and $\text{Re} < 5000$.

Clift et al. [7]

Clift et al. developed an improved expression which is valid for higher Re up to 3×10^5 and is as follows.

$$C_D = \frac{24}{R_e}\left(1 + 0.15\,\text{Re}^{0.687}\right) + \frac{0.42}{1 + 4.25\times10^4\,\text{Re}^{-1.16}} \qquad \text{--------9}$$

Garside and Al-Dibouni [10]

Garside and Al-Dibouni proposed a correlation for the multiparticle drag coefficient

$$C_D(\text{Re},\varepsilon) = \left[\frac{0.63}{u_{rs}} + \frac{4.8}{\sqrt{u_{rs}\,\text{Re}}}\right]^2 \qquad \text{--------10}$$

where u_{rs} is given by

$$u_{rs} = 0.5\left[A - 0.06\,\text{Re} + \sqrt{0.0036\,\text{Re}^2 + 0.12\,\text{Re}(2B - A) + A^2}\right]$$

Where $A = \varepsilon^{4.14}$

$B = 0.8\varepsilon^{1.28}$ $\quad \varepsilon \le 0.85$

$B = \varepsilon^{2.65}$ $\quad \varepsilon \ge 0.85$

Wan and Wu [34]

Gidaspow recommended the use of the following correlation for void fraction greater than 0.8

$$C_D = \frac{24}{\varepsilon R_e}\left(1 + 0.15(\varepsilon\,\text{Re})^{0.687}\right)\varepsilon^{-2.65} \qquad \text{--------11}$$

Comparison of predictions

The predictions from different correlations are presented in figures 4a, 4b and 4c. Figures 4a, 4b and 4c show the plot of calculated drag coefficient for a range of Reynolds number and for three values of sphericity, namely, ψ=0.6, 0.8 and 0.98 and for three different particle diameter, i.e., d_p=400, 500 and 650 μm. The accuracy of these widely used correlations are shown in figures 4a, 4b and 4c for a range of sphericity and particle diameter, while keeping the voidage constant without change. In depth analysis of existing predictive methods with particular importance to partial diameter, Reynolds number based on particle diameter, particle velocity, sphericity and void fraction are discussed.

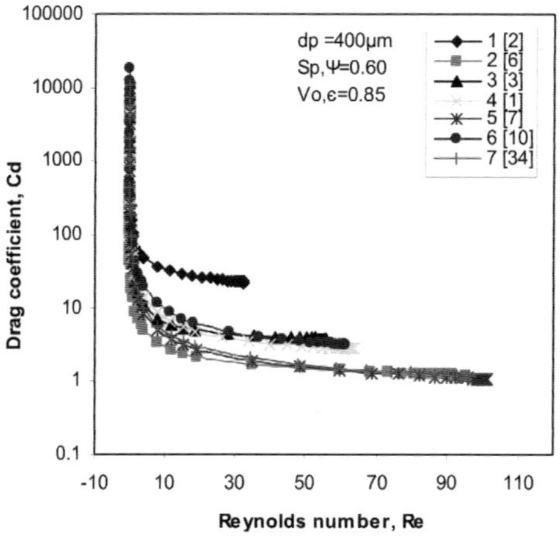

Figure 4 (a)

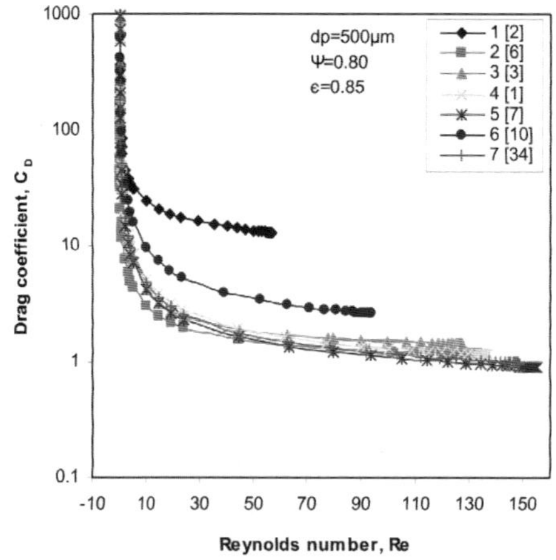

Figure 4 (b)

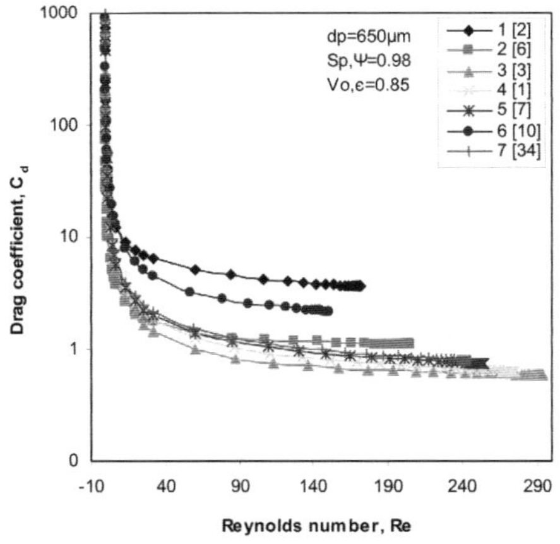

Figure 4 (c)

Figure 4. Comparison of drag coefficient predictions from existing methods for ψ=0.6-0.98; ε = 0.85; d_p=400 -650 μm

Ganser [2] and Garside and A-Dibouni [10] shows consistently a deviation from the rest and is shown in figure 4a-4c. The accuracy of Ganser [2] correlation may be due to its development based on particles experiencing a Stoke's regime where drag is linear to velocity and a Newton's regime where drag is proportional to the square of velocity. Also their correlation depends on two shape factors K_1 and K_2 which are unique functions of sphericity only. On the other hand Garside and A-Dibouni [10] correlation is good only when range of void fraction is available. This could be clearly seen in figure 7a. The estimation of the drag coefficient between Ganser [2] and Garside and A-Dibouni [10] has an error band of 45 % for Re in the range of 10^{-2} <Re<10^{1} and 67% for Re in the range of 10^{1} <Re<10^{2} over the entire range of particle diameter and sphericity.

Clift et al. [7] correlation is a simple and accurate and is valid for Reynolds numbers up to 3×10^{5}. This correlation is purely based on Re number only. It doesn't include any other parameter such sphericity or void fraction. Inspite of that, it shows consistent performance over the entire range, which can be visible in figure 4a-4c. Ganser [2] and Garside and A-Dibouni [10] over predicts Clift et al. [7] by 10% and 90% for Re in the range of 10^{-2} <Re10^{0}, whereas for Re in the range of 10^{0} <Re<10^{3}, 100-300% over prediction is seen. The rest of the prediction is within 30 %.

The correlation due to Swamee and Ojha [6] was developed based only on the data of Schul et al. [31], without accounting for the other literature data it also depends on the particle's shape factor called Corey shape factor which depends on the three principle axis of the particle. Accurate estimation of such parameter is difficult and may lead to a lower quality prediction. This can be seen in figures 4a-4c.
The relative performance of Chien [3], Haider and Levenspiel [1], Ganser [2] and Swamee and Ojha [6] can be noted in figure 4a-4c, as a function of sphericity.

Wen and Yu [34] and Garside and A-Dibouni [10] correlation was proposed for multiparticle drag coefficient. Garside and A-Dibouni [10] takes into account the effect of voidage and relative velocity between particle and fluid, whereas Wen and Yu [34] is simple and more reliable to use when void fraction plays a vital role in drag coefficient estimation. In the present analysis this correlation shows closer predictions after Clift et al. [5] to the data drawn from the literature.

The effect of particle diameter

Significant effect of particle diameter can be noted in figures 5a and 5b. Particle falling velocity is predicted through Eq.3 using individual drag relation listed from Eq. 4 to Eq. 11. The determination of falling and terminal velocity of the particle is of great interest in the design and optimization of solid particle receiver studies. *Particle Image Velocimetry (PIV)* will be used to trace the velocity of solid particles. PIV is a non-intrusive, full-field measurement technique, suitable for nominally two-dimensional flow fields, by which a velocity field is measured.

Figures 5a and 5b thus shows the velocity profile of the particle predicted by using Clift et al. [7] and Wen and Yu [34] drag correlation. For constant sphericity and void fraction the velocity profile shows significant variation for different particle diameter. The velocity of the particle increases with increase in particle diameter. The particle velocity increases and reaches a constant value once it attains its terminal velocity. The falling and terminal velocity can be noted in figure 5a and 5b. The lower the particle diameter, faster it can reach the terminal velocity.

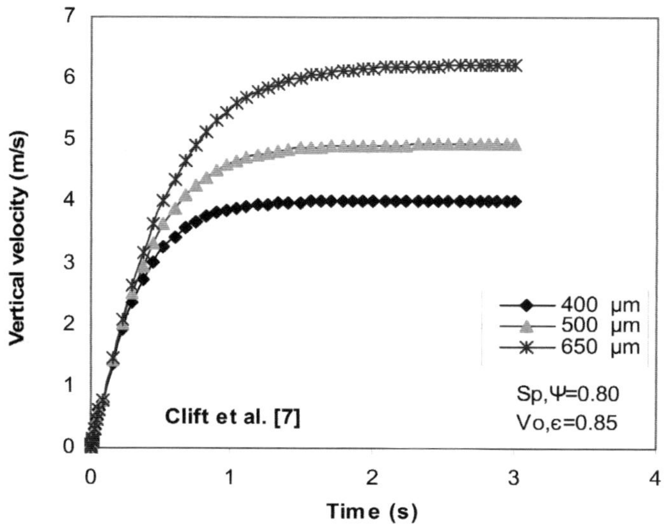

Figure 5 (a)

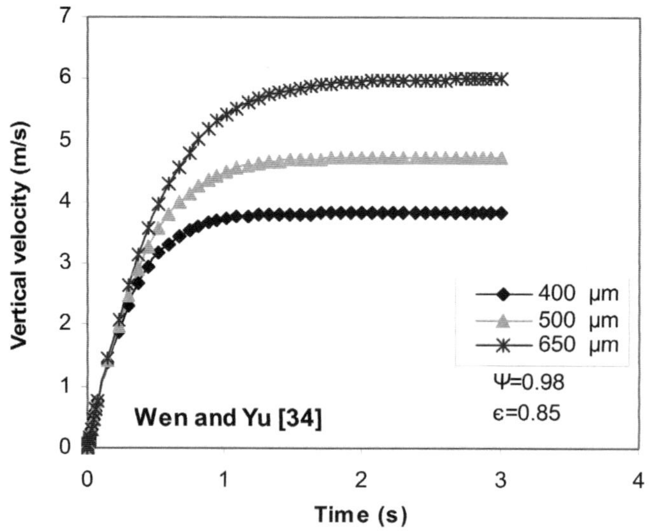

Figure 5 (b)

Figure 5. Prediction of particle falling velocity using existing drag relations for ψ=0.8; ε = 0.85; d_p=400 - 650 μm

The effect of sphericity

Sphericity, ψ, is defined as the ratio of the surface area of a sphere with equivalent volume to the actual surface area of the particle. For close to isometrically shaped particles, those with not very long or very short dimensions relative to other, the sphericity is a useful measure, most likely the best single parameter for describing the shape for falling particles. Figures 6a-6c shows the influence of sphericity for constant particle diameter and void fraction.

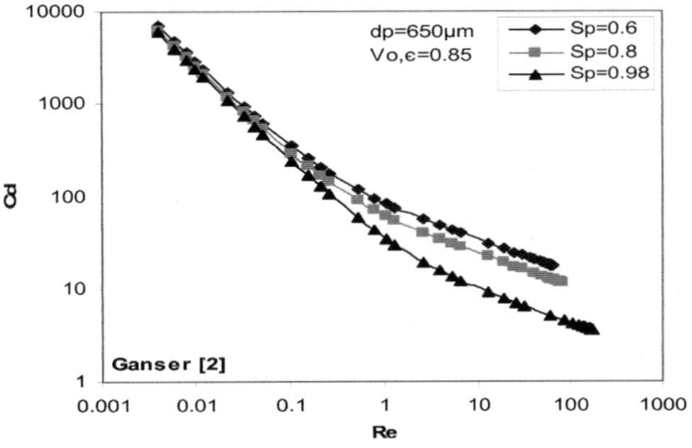

Figure 6 (a)

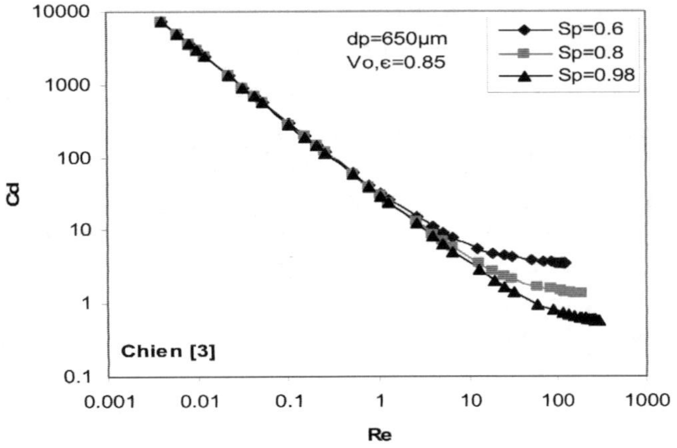

Figure 6 (b)

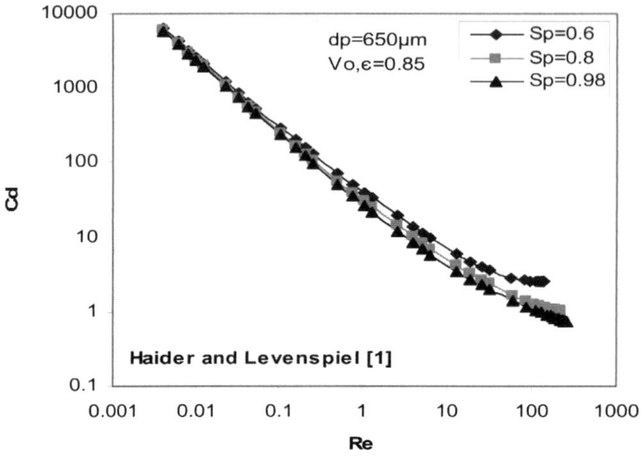

Figure 6 (c)

Figure 6. Effect of sphericity on drag coefficient for ψ=0.6-0.98; ε = 0.85; d_p=650 μm

Ganser [2], Chien [3] and Haider and Levenspiel [1] predictions are plotted in figures 6a-6c for sphericity ranging from 0.6 to 0.98. The strong influence of sphericity can be clearly seen in figures 6a-6c, as drag coefficient decreases with increase in sphericity. For spherical particles (ψ=1), the transition to settling velocity could be faster than for irregular particles.

The effect of voidage

Voidage/porosity is the volume fraction between the particles and is denoted as 'ϵ'. *Electrical-impedance tomography system* [28-29] could be used to measure solid concentrations 'ϵ_s, a noninvasive and with high temporal resolution. These methods require that the materials in the gas-solid flow should have distinct electrical properties. The measurement can be done invasively by inserting a probe fitted with electrodes into the flow or noninvasive by mounting electrodes on the boundary of the flow as shown in the figure 3. When multiple (i.e., more than two) electrodes are used, automated methods of data collection and mathematical reconstruction are needed. This approach is known most generally as electrical-impedance tomography (EIT). In electrical-impedance tomography, a number of electrodes are mounted to the surface of a domain of interest. As a prescribed current is injected into the domain at one electrode and withdrawn at another electrode as shown in figure 3, and all electrode voltages are measured relative to ground.

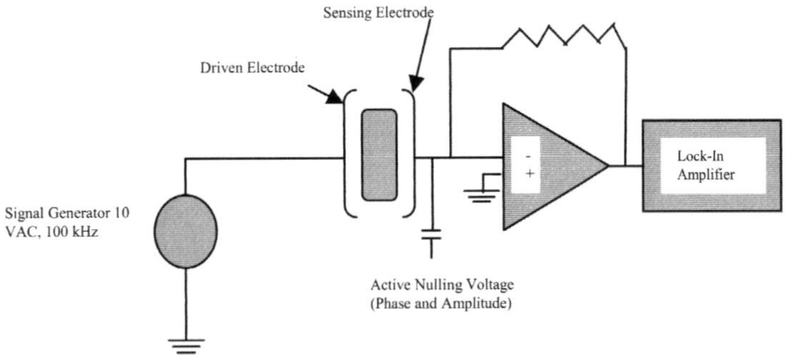

Figure 3. Bulk-impedance ring measurements circuit [28-29]

The plan is to use (EIT) the above method for the experimental determination of solid concentration and void fraction. Figure 7a and 7b shows the influence of void fraction on drag coefficient for constant particle diameter and sphericity. The results of Garside and A-Dibouni [10] and Wen and Yu [34] are plotted in figures 7a and 7b for void fractions ranging from 0.25 to 0.95. The transition to terminal velocity is much faster in the case of lower void fraction than in the case of higher void fraction. For example for void fraction 0.25 the particle reaches terminal velocity at time t = 0.21 seconds, whereas in the case of higher (0.95) void fraction the particle reaches a terminal velocity at time t = 1.50 seconds. Thus the measurement of void fraction is of importance in determining terminal velocity of falling particles.

Comparison of predictions with experimental data drawn from literature

Hruby et al. [33] performed experimental investigations with aluminum oxide particles as the working medium. They presented experimental data for both heated and unheated flow of solid particles. Instead of solar energy, they used electrical energy to heat the solid particles in the receiver hopper with band heaters ringed and controlled to maintain a constant interior wall temperature. To characterize particle behavior in a free fall of curtain of particles, they conducted some studies on unheated particle flow, which resulted in the measurement of particle vertical velocity with respect to fall height. They measured velocity of a single isolated spherical particle falling in quiescent surroundings using a TSI Laser Doppler Velocimetry system.

The data from their plot (variation of particle vertical velocity with fall height) is digitized and appropriate data corresponding to the present predictions are selected. The standard drag relation (Eq.2) is used to obtain the experimental drag coefficient.

Similar procedures are followed to draw experimental data from Turton and Levenspiel [35], Sabine Tran-Cong et al. [36] and Ceylan et al. [37] from the literature. The literature data are drawn for spherical particles having diameter and sphericity close to 650 μm and 0.98, respectively. Figure 8a shows the presentation of literature data as a function of Reynolds number.

The experimental data are compared with the aforementioned predictions as shown in figure 8. Since most of the experimental conditions lie for $Re > 10^1$, considerable importance have been given for $Re > 10^1$. However, the average relative error band is calculated for the entire range of Reynolds number. The average relative error in percentage variation for individual drag models with respect to the data drawn from the literature are presented in Table 2. A close scrutiny of results from Table 2 shows that Clift et al [5] predicts the data set with average overall error of 36.39 %. Even though the performance of Clift et al. [5] is appreciable over the entire range of Reynolds number, it does not include either void fraction or sphericity effects. While the relative error of Wen and Yu [34] is around an average overall error of 44.36 %, it is appropriate to point out that this correlation by Wen and Yu

[34] which is already a function of both Re number and void fraction and could be a suitable candidate, inspite of its deviation as compared to Clift et al. [5].

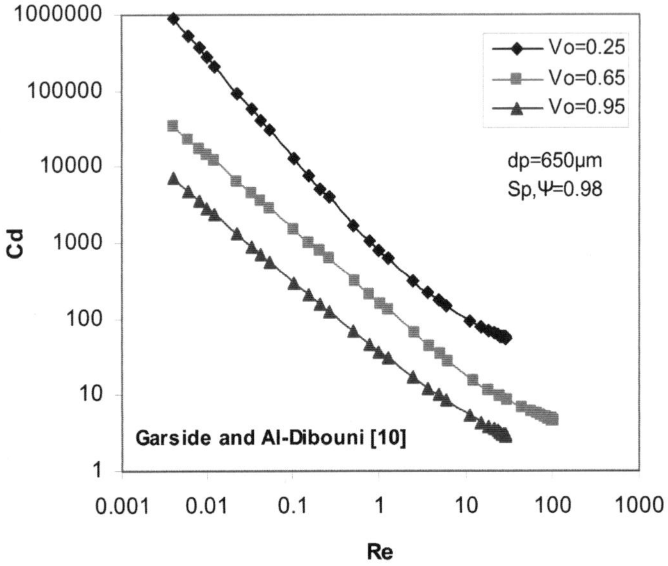

Figure 7 (a)

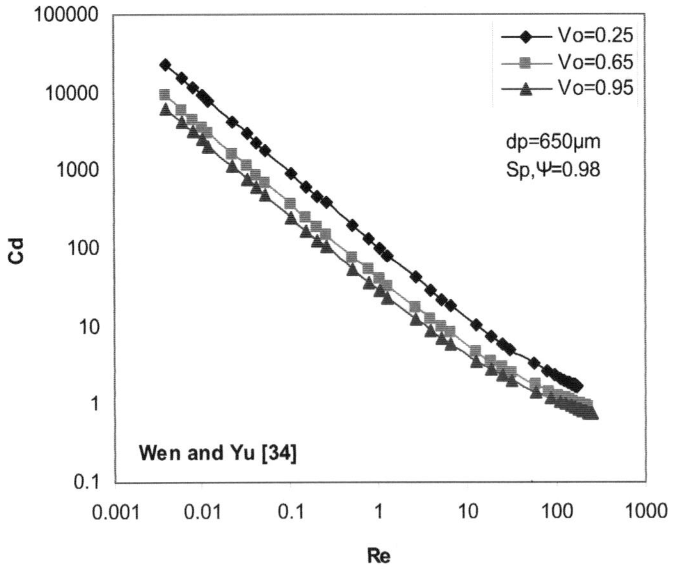

Figure 7 (b)

Figure 7. Effect of void fraction on drag coefficient for ψ=0.98; ε = 0.25-0.95; d_p=650 µm

Table 2. Percentage of deviation

Literature data	1 [2]	2 [6]	3 [3]	4 [1]	5 [7]	6 [10]	7 [34]	Proposed correlation
	Average relative error (%)							
[33]	273.60	42.61	71.08	56.83	37.65	32.29	30.13	18.90
[35]	56.88	51.66	19.21	17.21	15.73	15.92	43.1	9.61
[36]	75.80	23.84	34.9	52	28	68.55	46.81	21.34
[37]	76.29	267.81	71.7	73.73	64.2	81.22	57.42	27.34
Overall error (%)	120.64	96.48	49.22	49.944	36.39	49.49	44.36	19.29

Proposed correlation

The error analysis reveals the lack of a suitable drag relation that could predict the entire data sets by considering the influence of all important parameters. This suggests a basis to derive an improved drag correlation which could overcome this need in the literature. Most of the existing drag correlations were developed with Reynolds number as a primary and important function. Very few though are available which include void fraction or sphericity as a co-function of Reynolds number. This could be further justified from the above discussion in that the effects of particle diameter, void fraction and sphericity have considerable impact on drag coefficient and clearly, a drag correlation by incorporating these parameters would fill the lack of information in the literature. Hence, a suggested drag coefficient model incorporating Re number, as well as particle shape, ψ, particle diameter, d_P and void fraction $\in$ is proposed here which could be represented in the form of

$$C_D = C_D(\text{Re}, \psi, d_P, \in) \qquad \text{--------12}$$

In this study Wen and Yu [34] shows better agreement and thus the modified correlation takes the following form with the inclusion of parameters represented in Eq.12

$$C_D(R_e, \varepsilon, \psi) = \frac{24}{\varepsilon R_e}\left(1 + 0.15\left(\varepsilon \,\text{Re}^{0.587}\right)\right)\varepsilon^{-2.25} + (\psi R_e)^{0.35} \exp(-5.05\psi) \qquad \text{--------13}$$

In Eq.13 the additive term, product and exponents were obtained using regression analysis. The performance of the above correlation and its better prediction of 242 literature data within about an average overall error of 19.29% is shown in figure 8. The proposed correlation in Eq.13 is valid for $10^0<\text{Re}<10^4$; particle diameter ranging from 100 μm to 650 μm; void fraction from 0.2 to 0.95 and sphericity from 0.4 to 0.98.

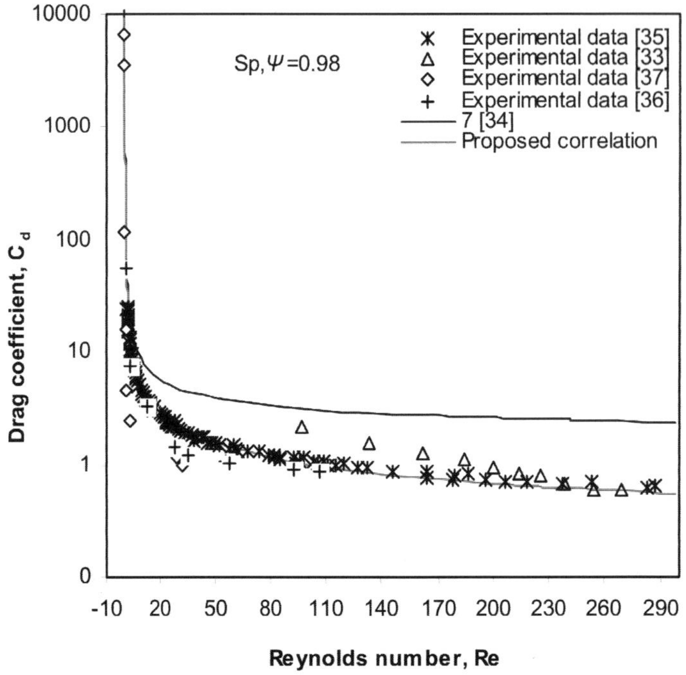

Figure 8. Performance of the proposed correlation with four sets of experimental data drawn from the literature

The strength of the proposed correlation (Eq.13) is further justified by testing the correlation for individual effects such as void fraction and sphericity with the experimental data drawn from the literature.

Haider and Levenspiel [1] have used experimental data compiled from Pettyjohn and Christiansen [40] for different particle sphericities including isometric and non-isometric shapes. The data of Pettyjohn and Christiansen [40] presented by Haider and Levenspiel [1] were drawn for Ψ = 0.670 and 0.906. Figure 9 shows the performance of the proposed correlation for particle sphericity of 0.670 and 0.906. The correlation predicts the entire data set within an average relative error of 13.71%. Panigrahi and Murty [38] tested their model by using the fluidized bed data from Barnea and Mizrani [39]. In the case of fluidized beds, the void fraction between the fluid and the particles increases and reaches unity when the bed is fully expanded, which means that the particles suspended in the fluid behave like single isolated particles. Thus, the drag data for different void fraction presented by Panigrahi and Murty [38] from the velocity data of Barnea and Mizrani [39] were drawn for comparison with the proposed correlation. The performance of the correlation is shown in figure 10, with the predictions depicting a relative error of 22.72% for spherical particles with void fraction, ϵ of 0.45 and 0.75.

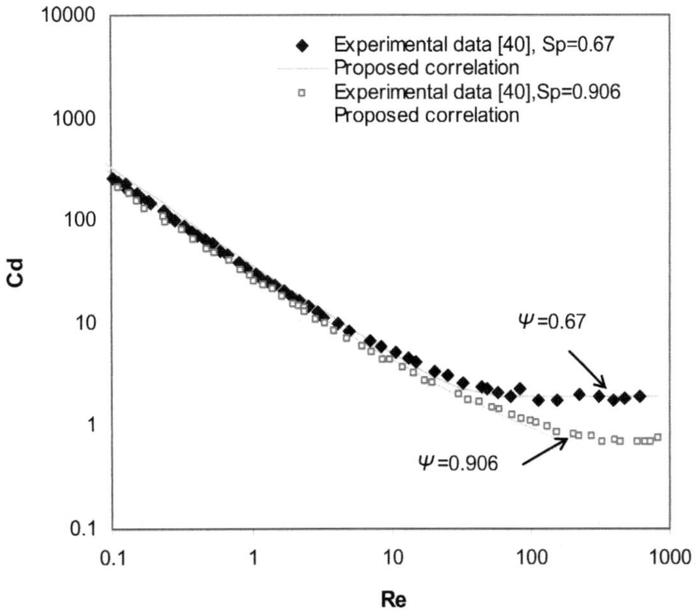

Figure 9. Performance of the proposed correlation with the data of Pettyjohn and Christiansen [40] for sphericities, $\psi = 0.67$ and 0.906

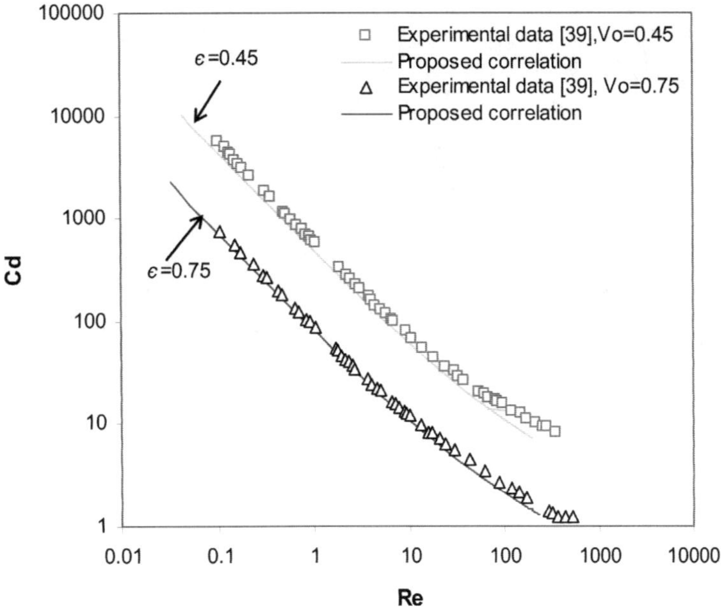

Figure 10. Performance of the proposed correlation with the data of Barnea and Mizrani [39] for void fraction, $\varepsilon = 0.45$ and 0.75

Conclusion

As part of solid particle receiver studies, a fundamental understanding of particle behavior is necessary for further design of receiver. This work presents the analysis of widely used drag models/correlations with particular reference to solid particle of spherical shape. Importance of particle diameter, void fraction and sphericity and its influence on particle velocity and terminal velocity are also discussed. A correlation is proposed based on Wen and Yu by incorporating the influential parameters of both void fraction and sphericity. The proposed correlation predicts the four sets of literature data to within the overall error of 19.29%. The strength of the correlation is further proved by testing the correlation with the additional experimental data drawn from the literature separately for void fraction and sphericity.

Acknowledgement

The authors wish to thank Gregory J. Kolb and Nathan Siegel, National Solar Thermal Test Facility, Sandia National Laboratories, Albuquerque, NM, USA, for their facilities and help.

Nomenclature

A	Cross-sectional area of the spherical particle
C_d	Drag coefficient
d_p	Particle diameter, m
g	Acceleration due to gravity, m/s^2
m_p	Mass of the spherical particle, kg
Re	Reynolds number
u_p	Particle velocity , m/s
u_{rs}	Relative velocity , m/s
V	Particle volume

Greek symbols

β	Corry shape factor
ρ_s	Particle density, kg/m^2
ρ_f	Fluid density, kg/m^2
ψ	Sphericity
ε	Void fraction

References

[1] Haider, A. M., Levenspiel, O., 1989, "Drag Coefficient and Terminal Velocity of Spherical and Nonspherical Particles," Powder Technology, 58, pp.63-70.

[2] Ganser, G. H., 1993, "A Rational Approach to Drag Prediction of Spherical and Nonspherical Particles," Powder Technol., 77, pp.143-152.

[3] Chien, S. F., 1994, "Settling Velocity of Irregularly Shaped Particles," SPE Drilling and Completion, 9, pp.281.

[4] Hartman, M., Trnka, O. and Svoboda, K., 1994, "Free Settling of Nonspherical Particles," Ind. Eng. Chem. Res., 33, pp.1979.

[5] Turton, R., Levenspiel, O., 1986, "A Short Note on the Drag Correlation for Spheres," Powder Technol., 47, pp.83-86.

[6] Swamee, P. K., Ojha, C. P., 1991, "Drag Coefficient and Fall Velocity of Nonspherical Particles," J. Hyd. Eng., 117, pp.660-667.

[7] Clift, R., Grace, J. R., Weber, M.E., 1978, "Drops and Particles'', Academic Press, New York.

[8] Sabine Trang-Cong, Michael Gray, Efstathios E. Michaelides, 2004, "Drag Coefficients of Irregularly Shaped Particles," Powder Technol. 139, pp.21-32.

[9] Syamlal, M., O'Brien, T. J., 1987, "The Derivation of a Drag Coefficient Formula from Velocity-Voidage Correlations," The unpublished article.

[10] Garside, J., Al-Dibouni, M. R., 1977, "Velocity-Voidage Relationship for Fluidization and Sedimentation in Solid-Liquid Systems," Ind. Eng. Chem., Process Dev., 16, pp.206-214.

[11] Steinfeld, A., Imhof, A., and Mischler, D., 1992, "Experimental Investigation of an Atmospheric-Open Cyclone Solar Reactor for Solid-Gas Therochemical Reactions," ASME Journal of Solar Energy Engineering, 114, pp.171-174.

[12] Steinfeld, A., Frei, A., Kuhn, P., and Wuillemin, D., 1995, "Solar Thermal Production of Zinc and Syngas via Combined ZnO-Reduction and CH4-Reforming Processes," International Journal of Hydrogen Energy, 20, pp.793-804.

[13] Steinfeld, A., Frei, A., Kuhn, P., and Wuillemin, D., 1998, "Research and Development of the Process Technology for Converting Concentrated Solar Energy into Chemical Fuels," PSI—EF-REN (92) 033.

[14] Haueter,P., Moeller, S., Palumbo, R., and Steinfeld A., 1999, "The Production of Zinc by Thermal Dissociation of Zinc Oxide Solar Chemical Reactor Design," Solar energy, 67, pp.161-167.

[15] Abdelrahman, P., Fumeaux, P., and Suter, P., 1979, "Study of Solid–Gas Suspension Used for Direct Absorption of Concentrated Solar Radiation," Solar Energy, 22, pp.45-48.

[16] Hunt, A.J., 1979, "A New Solar Receiver Utilizing a Small Particle Heat Exchanger," Proceedings of the 14th International Society of Energy Conversion Engineering Conference, New York: Institute of Electrical and Electronics Engineers; pp. 159-163.

[17] Hunt, A.J., and Brown, C.T., 1982, "Solar Testing of the Small Particle Heat Exchanger (SPHER)," Berkley (CA): Lawrence Berkeley National laboratory, Report no. LBL -16497.

[18] Ganz, J., Schelling, T.h., and Sturzenegger, M., 1997, "A New Design for Volumetric Gas/Particle Reactors for the Solar Thermal Reduction of Metal Oxides," PSI Annual report 1997/ Annex V, pp.8-9

[19] Hruby, J.M., "A Technical Feasibility Study of a Solid Particle Solar Central Receiver for High Temperature Applications," Sandia National Laboratories, SAND 86-8211.

[20] Martin, J. and Vitko, J., 1982, "ASCUAS: A Solar Central Receiver Utilizing a Solid Thermal Carrier," Sandia national Laboratories, SAND 82-8203.

[21] Falcone, P.K., 1984, "Technical Review of the Solid Particle Receiver Program," Sandia National Laboratories," SAND 84-8229

[22] Sargent and Lundy, 2003, "Assessment of Parabolic Trough and Power Tower Solar Technology Cost and Performance Forecasts," SL-5641

[23] Pohl, P.I., Brown, L.C., Chen, Y., Diver, R.B., 2004, "Evaluation of Solar Thermo chemical Reactions for Hydrogen Production," Proceedings of the 12 th International Symposium Solar Power and Chemical Energy Systems.

[24] Westinghouse Electric Corp., 1982, "Solar Thermal Hydrogen Production Process," DOE/ET//20608-1.

[25] Hilderbrandt, A.F., and Ross, K.A., 1985, "Receiver Design Considerations for Solar Central Receiver Hydrogen Production," 35, pp.199-206.

[26] Gregory, J. Kolb, R.B.Diver and Nathan Siegel, 2005, "Central-Station Solar Hydrogen power Plant," Proceedings of ISEC, Orlando, Florida, USA.

[27] Robert S. Okojie, Gustave Fralick, and George J. Saad, "SiC-Based Miniature High-Temperature Cantilever Anemometer," John H. Glenn Research Center, Cleveland, Ohio.

[28] George, D. L., Torczynski, J. R., Shollenberger, K. A., O'Hern, T. J., and Ceccio, S. L., 2000, "Validation of electrical-impedance tomography for measurements of material distribution in two-phase flows," International Journal of Multiphase Flow, 26, pp. 549-581.

[29] Tortora, P. R., Ceccio, S. L., Trujillo, S. M., O'Hern, T. J., and Shollenberger, K. A., 2004, "Capacitance measurements of solid concentration in gas-solid flows", Powder Technology, 148, pp. 92-101.

[30] James O Wilkes, 1999, "Fluid Mechanics for Chemical Engineers," Prentice Hall International Series.

[31] Schul, E.F., Wilde, R.H., Albertson, M.L., 1954, Corps of Engineers, U.S. Army, through Colarado A&M Research Foundation, Fort Collins, CO.

[32] Huajan Chen, Yitung Chen, Hsuan-Tsung Hsieh and Nathan Siegel, 2006, "CFD Modeling of Gas Particle Flow within a Solid Particle Solar Receiver," Proceedings of ISEC, Denver, USA.

[33] Hruby, J.M., Steeper, R.R., Evans, G.H and Crowe, C.T. "An Experimental and Numerical Study of Flow and Convective Heat Transfer in a Freely Falling Curtain of Particles," Sandia National Laboratories, SAND 86-8714.

[34] Wen, C.Y., and Yu, Y.H., 1966, "Mechanics of Fluidization '', Chemical Engineering Progress Symposium Ser., 62, pp. 100-111.

[35] Turton, R., and Levenspiel, O., 1986, "A Short Note on the Drag Correlation for Spheres", Powder Technology, 47, pp. 83-86.

[36] Sabine Tran-Cong, Michael Gay, Efstathios E. Michaelides, 2004, "Drag Coefficients of Irregularly Shaped Particles", Powder Technology, 139, pp. 21-32.

[37] Ceylan, K., Herdem, S., and Abbasov, T., 1999, "A Theoretical Model for Estimation of Drag Force in the Flow of Non-Newtonian Fluids around Spherical Solid Particles", Powder Technology, 103, pp. 286-291.

[38] Panigrahi, M.R. and Murty, J.S., 1991, "A Generalized Spherical Multiparticle Model for Particulate Systems: Fixed and Fluidized", Chemical Engineering Science, 46, pp. 1863-1868.

[39] Barnea, E. and Mizrani, J., 1973, "Generalized Approach to the Fluid Dynamics of Particulate Systems. Part I General Correlation for Fluidization and Sedimentation in Solid Multiparticle Systems", Chemical Engineering Journal, 5, pp. 171-189.

[40] Pettyjohn, E.S. and Christiansen, E.B., 1948, "Effect of Particle Shape on Free-Settling Rates of Particles of another 10 Patients served", Chemical Engineering, 2950, pp. 157-172.

431a Stability Of Cationic Headgroups In Alkaline Anion-Exchange Membrane Fuel Cells

Shaji Chempath[1], Brian R. Einsla[2], James M. Boncella[3], Bryan R. Pivovar[2], and Lawrence R. Pratt[1]. (1) T-12, Mailstop B 268, Los Alamos National Lab, Los Alamos, NM 87545, (2) MPA-11, Los Alamos National Lab, Los Alamos, NM 87545, (3) MPA-MC, Los Alamos National Lab, Los Alamos, NM 87545

Alkaline anion-exchange membranes (AAEM) can be used as the polymer electrolyte in fuel cells which operate under alkaline conditions [1]. These membranes have cationic headgroups (e.g., [N(CH3)4]+) whereas the commonly used acidic membranes have an anionic headgroup (e.g., HSO3- in Nafion). Alkaline conditions allows the use of catalysts without platinum and may provide better water management properties. However, currently available AAEM materials are chemically less stable compared to their acidic counterparts such as Nafion. The headgroups of AAEMs are chemically attacked by OH- and undergoes degradation leading to loss of conductivity under the operating conditions in a fuel cell. Density functional theory (DFT) calculations reveal the critical role played by water in these membrane degradation reactions.

Electronic structure calculations with a polarizable continuum model (Gaussian 03 software,B3LYP/6-311++g(2d,p) level theory) of aqueous tetramethylammonium ([N(CH3)4]+) cations revealed two mechanisms of degradation. In the first mechanism, the OH- attacks a methyl group on the [N(CH3)4]+ in SN2 fashion and forms CH3OH and trimethylamine (N(CH3)3). In the second mechanism the OH- extracts a proton from a CH3 group and leads to the formation of water and a ylide - trimethyl ammonium methylide ((CH3)3 N+ (CH2)-). This ylide complex then undergoes additional degradation and forms methanol and trimethylamine. Reaction paths for both these mechanisms were obtained with the growing string method [2]. The highest saddle point in both mechanisms are equivalent and the activation barrier is +13.5 kcal/mol. Experimental measurements confirms (Mass Spectroscopy and Thermogravimetric Analysis) that CH3OH and trimethylamine are formed during the degradation of the model compound tetramethylammonium hydroxide pentahydrate (N(CH3)4OH-(H2O)5) . Mass spectrometry during the degradation of the deuterated version ((CH3)4N][OD](D2O)5) shows the scrambling of D and H on the methyl groups and thus confirms the presence of the ylide mechanism.

DFT calculations were repeated by changing the di-electric constant of the polarizable continuum medium. It was found that OH- becomes more reactive (i.e., smaller activation barrier) when the dielectric constant is reduced. This is expected because OH- will be poorly solvated at low dielectric constants.

Ab-initio MD simulations (VASP software, PW91 exchange) with one tetramethylammonium, one OH-, and n water molecules (n = 0 .. 4) showed that the head group is attacked by hydroxide when n=0 or n=1. However when four water molecules are present in the system the OH- is well solvated by the water molecules and no reaction is observed. Ab-initio MD simulations in a periodic cell with 48 water molecules also show no reaction. Radial distribution functions obtained from such simulations show that hydroxide is not tightly bound near the TMA+ cation. This is because the +1 charge of tetramethylammonium is spread uniformly around the bulky cation.

These results imply that a OH- anion that is not well hydrated will attack the cationic head groups in the AAEM. An alkaline membrane operating under relatively dry conditions will degrade much faster compared to a well hydrated membrane.

REFERENCES

[1] Prospects for alkaline anion-exchange membranes in low temperature fuel cells.; Varcoe, JR and Slade, RCT; Fuel Cells; 2005 (5), pp 187-200.

[2] A growing string method for determining transition states: comparison to the nudged elastic band and string methods.; Peters B, Heyden A, Bell AT, and Chakraborty A; J. Chem. Phys.; 2004 (120), pp 7877-7886.

431b Molecular Dynamics Simulations Of Nanoscale Hydrophilic/ hydrophobic Domain Structure In Hydrated Nafion

Craig K. Knox[1], Matthew K. Petersen[2], and Gregory A. Voth[2]. (1) Department of Chemical Engineering, Center for Biophysical Modeling and Simulation, University of Utah, Salt Lake City, UT 84112, (2) Department of Chemistry, Center for Biophysical Modeling and Simulation, University of Utah, Salt Lake City, UT 84112

Experimental elucidation of the nanoscale structure of hydrated Nafion, the most popular polymer electrolyte membrane (PEM) to date, and its influence on macroscopic proton conductance is particularly challenging. While it is generally agreed that hydrated Nafion is organized into distinct hydrophilic/hydrophobic domains, the geometry and length scale of these domains continues to be debated. For example, at least half a dozen different domain shapes, ranging from spheres to cylinders, have been proposed based on experimental SAXS and SANS studies. Since the characteristic length scale of these domains is believed to be ~2 to 5 nm, which is approximately equal to the box size of a typical ~5,000 atom PEM system, large atomistic or coarse-grained molecular dynamics (MD) simulations are needed to accurately probe the structure and morphology of these domains, especially their connectivity and percolation phenomena. Using classical, atomistic MD with explicit hydronium ions and a flexible water molecule, we have simulated a very large hydrated Nafion system (~2.3 million atoms in a cube with a side length of ~30 nm) at varying water contents to directly observe several hydrophilic domains at the molecular level and to probe the morphology and lifetimes of the connecting "structures" between them. This talk will explore the results of these simulations, which may help provide guidance for the design of new alternative membranes with superior proton conductance that will improve overall PEM fuel cell performance.

431c Reactive Molecular Dynamics Applied To Proton Transport In Fuel Cells

Myvizhi Esai Selvan[1], Junwu Liu[1], David J. Keffer[1], Brian J. Edwards[1], Shengting Cui[1], and William V. Steele[2]. (1) Department of Chemical Engineering, University of Tennessee, Knoxville, TN 37996, (2) Nuclear Science and Technology Division, Oak Ridge National Laboratory (ORNL), Oak Ridge, TN 37831-6181

Our goal is to develop and apply a coarse-grained description of proton transport that is based on quantum mechanical calculations and correctly reproduces the macroscopic reaction rates. Our approach is to modify the classical equilibrium molecular dynamics simulation algorithm using existing non-reactive potentials. The advantage of this is that we use a computationally efficient, non-reactive potential that already exists and have been optimized. For water, we use the TIP3P potential, which does not explicitly allow a hydronium ion to transfer a proton to a water molecule. The reactive algorithm has three steps. The first step is the reaction trigger. We examine every hydronium ion in the simulation at each time step to determine whether it is in a state where it would transfer a proton. The determination of this state is based upon a set of reaction triggers, some of which are geometric and others energetic. The geometric triggers are based upon quantum mechanical configurations. The energetic limit is tied to the activation energy. If a hydronium ion satisfies all of the geometric and energetic triggers, the reaction instantly takes place. The second step of the reactive algorithm is the instantaneous reaction. In this case, the reaction involves replacing the bond-stretching (harmonic) interaction of the O of the reactant H_3O^+ and the proton to be transferred with a non-bonded (Lennard-Jones) interaction. The reverse switch is made for the reactant H_2O and the proton to be transferred. The proton to be transferred is moved along the OO axis such that the ratio of O^*-H (where O^* is in H_3O^+) and O-H distances are the same in the product molecules as they were in the reactant molecules. The third and final step of this procedure is local equilibration, which is performed to exactly satisfy the target, ΔH_r heat of the reaction. In this reaction, ΔH_r, is zero because the reactants and products of the reaction are identical. Once all three steps of the reactive algorithm are complete, we continue the classical simulation using the non-reactive potential. Using this procedure, we can match all four elements that uniquely define a macroscopic description of the reaction: the stoichiometry, frequency prefactor k_o in the rate constant, activation energy E_a and ΔH_r. The matching of the stoichiometry is mandated through the reaction trigger. We can adjust the numerical values of the limits on the geometric triggers to control k_o and the limit on the energetic trigger to control E_a. We have developed and implemented this algorithm, which is simple and efficient. It automatically responds to the environment in the bulk water and hydrated Nafion in the fuel cell.

Acknowledgments: The work is supported by a grant from the U. S. Department of Energy BES under the contract number DE-FG02-05ER15723. This research used resources of the Center for

Computational Sciences at Oak Ridge National Laboratory, which is supported by the Office of Science of the DOE under Contract DE-AC05-00OR22725.

431d Method For Simulating Proton Transport

Chun-Min Chang and **Dean Wheeler**. Chemical Engineering Department, Brigham Young University, Provo, UT 84602

An improved understanding of proton-transport mechanisms in aqueous and hydrophilic systems such as low-temperature fuel cells can be helpful in optimizing such systems. In hydrogen-bonded networks, protons transport by both vehicular and structural diffusion. In particular, structural diffusion which involves the making and breaking of covalent bonds is difficult to model efficiently in large-scale atomistic simulations.

We present here a model for simulating proton structural diffusion in molecular dynamics codes. The model uses charge-equilibration techniques to better describe behavior in heterogeneous environments. The quantum aspects of making and breaking covalent bonds are treated with an empirical valence bond (EVB) approach combined with the principles of Car-Parrinello molecular dynamics. This work demonstrates the nature of collective proton conduction in water under different conditions. Future optimization and improvement of the model are also discussed.

431e Molecular Dynamics Simulations Of Lithium Transport In Bulk Electrolytes And At The Electrolyte/electrode Interface

Oleg Borodin, Materials Science and Engineering, University of Utah, 122 S. Central Campus Dr. Rm 304, Salt Lake City, UT 84112 and Grant D. Smith, Materials Science & Eng., University of Utah, 122 S. Central Campus Dr. Rm 304, Salt Lake City, UT 84112.

Gaining fundamental understanding of mechanisms of ion transport is important for design of novel electrolytes with improved conductivity. Molecular dynamics (MD) simulations are well suited for exploring transport mechanisms on nanosecond time scales. In this presentation we focus on utilizing MD simulations for exploring mechanisms of ion transport in a number of liquid electrolytes and ionic liquids as well as the electrolyte structure next to electrodes and the SEI layer. The free energy barrier associated with the lithium transport from the electrolytes into electrolytes has been obtained from MD simulations. It was found is good agreement with the to the experimental activation energies for the charge transfer for similar liquid electrolytes. Dependence of the free energy barrier for the lithium transfer from electrolytes into electrodes on the electrolyte composition has been studied using MD simulations and will be discussed.

487a Molecular Modeling Of Anhydrous Polymeric Membranes For Fuel Cells

Stephen J. Paddison, Chemistry and Materials science, University of Alabama in Huntsville, 301 Sparkman Dr., Huntsville, AL 35899

The molecular features governing the transfer of protons in recently synthesized membranes prepared from polymeric organic amine salts are investigated with first principles based electronic structure calculations. Global minimum energy structures were determined for bisulfates and dihydrogenphosphates of monomeric fragments of 2-vinylpyridine, 4-vinylpyridine, and vinylimidazoline. The equilibrium structures fully optimized at the B3LYP/6-311G** level indicate that the energetically preferred state for the phosphates is consistently as a phosphoric acid/base pair in contrast to the bisulfates of 2-vinylpyridine and vinylimidazoline where a proton is transferred to the base resulting in the protonated base/bisulfate contact ion pair. Binding energies of the acid/base or anion/protonated base pairs were computed from equilibrium structures fully optimized with the counterpoise method to correct for basis set superposition error and show that the transfer of the proton results in substantially stronger binding. The energetics for the transfer of the proton from the acid to the amine base and revealed that if the transfer is constrained (i.e. the O–H…N hydrogen bond) a barrier of as much 2.5 kcal/mol occurs. However, if the transfer is performed without imposing any constrains on either molecule the barrier disappears. Proton transfer with phosphoric acid showed endothermic penalties of 6.0, 5.0 and 2.2 kcal/mol for 4-vinylpyridine, 2-vinylpyridine, and vinylimidazoline, respectively. This was in contrast to sulphuric acid where the penalties were exothermic by 0.8 and 4.4 kcal/mol for 2-vinylpyridine and vinylimidazoline, but only very slightly endothermic with 4-vinylpyridine. These energetics are generally consistent with the low temperature conductivities of solution-cast polymeric amine salts determined from impedance measurements. Furthermore, these results imply that the measured high temperature (i.e. > 120 °C) conductivities are not simply due to the hopping of protons but involve the mobility of the larger protonic charge carriers (i.e. phosphoric acid).

Characterization Of Mediators Proposed To Better Facilitate Electron Transfer In Methanol Dehydrogenase Enzymatic Fuel Cells

Katherine L. Keeton and Daniela S. Mainardi
Louisiana Tech University, Ruston, Louisiana

Abstract

Methanol Dehydrogenase enzyme (MDH) is considered for bio-electrochemical generation of power as a result of methanol oxidation. The use of an artificial electron mediator is being investigated in this work using quantum mechanics and molecular mechanics simulations to see if more effective transferring of the electron from the enzyme to the substrate can occur.

Introduction

The MDH enzyme is readily available in nature in soil and other parts of plants and has the ability to oxidize methanol and other primary alcohols to their corresponding aldehydes. Its crystal structure from M*ethlylobacterium extorquens 1H4I* [1] and from *Methylophilus W3A1* [2] has been characterized using x-ray crystallographic methods [1-4].

The complete MDH contains over 24,000 atoms in two subunits consisting of two lobes, each containing an active site, which is responsible for the catalytic behavior of the enzyme. At the center of the active site is a calcium ion with a +2 charge. It is encased by a Pyrroloquinoline quinone (PQQ) molecule that is a cofactor required by the enzyme [5-6]. Thirteen amino acids surround the PQQ molecule to complete the active site as shown in Figure 1.

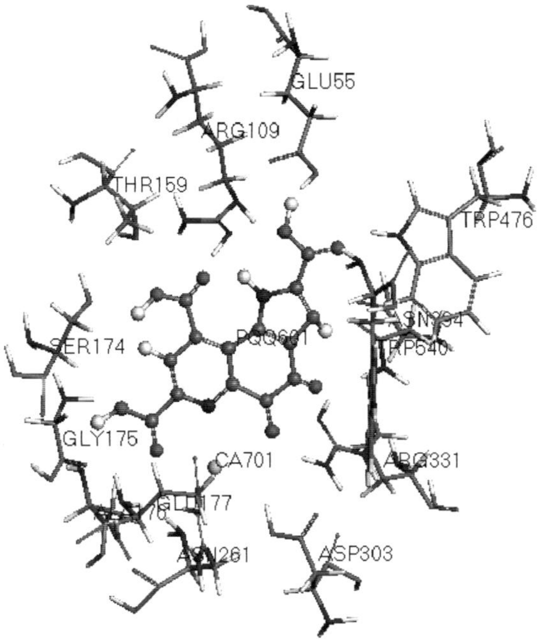

Figure 1. Active site of Methanol Dehydrogenase enzyme (MDH). Information obtained from *Methylobacterium Extorquens,* entry 1H41 of the Protein Fata Bank. [7]

The MDH active center is located deep within each of the lobes of the MDH enzyme. It is believed that this location obstructs efficient electron transfer from the active site to the electrode supports and literature suggests that the remaining amino acids present in the enzyme provide insulation and prohibit electron transfer [8]. MDH is naturally attached to an electron mediator, cytochrome cL. The effectiveness of this molecule as an electron transporter has been studied [9] but the use of artificial electron transfer mediators has been proposed to be more beneficial in providing efficient electron transport [10].

A prototype of a biocatalytic fuel cell has been constructed and it utilizes MDH as the catalyst with methanol as a fuel source,[11] and considers a solution of an artificial mediator, N,N,N,N-tetramethyl-phenylenediamine (TMPD). It was proposed that the cell would have a theoretical continuous power output of 100mW/cm^2 for at least 30 days of continuous operation [10]. However, the reported actual cell power output turned out to be 0.25 mW/cm^2 for two weeks of continuous operation [11]. The authors observed 0.25% of the targeted output, indicating that there are limiting factors. Biological fuel cells, in general, have the efficiency of approximately 40%, compared to that of chemical fuel cells, which have efficiencies ranging from 40%-60%[12]. One of the limiting factors proposed for the previously mentioned experiment was the ability of the electron transfer from the active site [10]. Molecules of different composition and size lend promise to finding more efficient means of electron transport to produce larger power outputs.

In this work two molecules have been investigated as possible mediators for the methanol oxidation by MDH due to their electron transfer abilities [13]. In addition to TMPD, Tris-(methoxy)carboxylethylsilane (TMCES), is investigated and their abilities to provide efficient electron transport in MDH evaluated. The suggestion that there must be a more appropriate mediator can be determined by the comparison of the different attributes of these molecules that may improve the facilitation of electron transport. In order to facilitate electron transport the molecule must be able to attract the electron, move the electron through the molecule, and then release the electron towards the electrode support. The last described step may release the electron through another molecule of the proposed mediator or to an amino acid contained within the MDH enzyme, where the movement is away from the active site center. These proposed movements for an electron can be characterized by the ionization potential and electron affinity of the mediator molecule.

Methodology

In this work a variety of quantum mechanical methods have been applied. Information about the geometric and energetic properties of the mediators has been obtained from Ab Initio and Density Functional Theory (DFT) simulations. Using Gaussian '03 software, Simple Hartree Fock (HF) computations were first examined. The computational efficiency of these simple calculations provided the ground work for the investigation. Then DFT, which incorporates electron correlation, was used to refine the calculations [14]. Two different DFT theory levels were examined considering hybrid and Generalized Gradient Approximation (GGA) methods. The hybrid method Becke 3 Perdew-Wang 91(B3PW91) was used with 6-311+G** as the basis set, which is a combination of split-valence and triple zeta basis functions. The GGA method used in this work was the PW91 functional in combination with the DNP basis set.

Molecular mechanics calculations were further performed using the Condensed-phase Optimized Molecular Potentials for Atomistic Simulation Studies (COMPASS) forcefield to find energy minimized structures for larger systems. COMPASS is a powerful forcefield because it is the first ab inito forcefield that enables accurate prediction of gas-phase and condensed-phase properties. It also has the ability to consolidate parameters of organic and inorganic materials [15].

Results

Modeling the MDH Enzyme

In order to validate the different theory levels selected, calculated optimized structural parameters were compared to reported experimental information. The distances between selected atoms and the calcium ion were compared and are shown in Table I. The computational calculations were taken from HF and DFT simulations. The experimental data was obtained from X-ray crystallography [3,5,9].

There is a small difference between the computational DFT (GGA) and those found experimentally by Ghosh et al.[9]. All of the computationally found distances between the molecules shown are within acceptable range, taking into account that the two techniques for acquiring the data are completely different. The complimentary data shown in this baseline calculation of the center part of the active site validates the level of theory chosen.

	Method	Ca2+-N	Ca2+-O1	Ca2+-O4
Computational	HF	2.46	2.30	2.36
	DFT-Hybrid	2.40	2.30	2.29
	DFT-GGA	2.45	2.77	2.52
X-ray crystallography	1.2 Å (Williams et al.)[3]	2.32	2.25	2.44
	1.9 Å (Xia et al.)[5]	2.4	2.5	2.44
	1.94 Å (Ghosh et al.)[9]	2.45	2.77	2.44

Table I. Comparison of distances (Angstroms) between atoms found computationally at different theory levels versus distances found at different depths of x-ray crystallography for PQQ and Ca^{2+} ion.

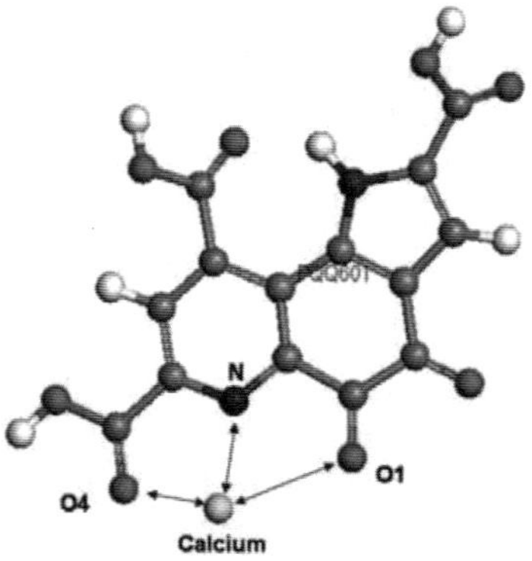

Figure 2. Indicated lengths from PQQ and calcium ion.

Mediators

TMPD is a symmetric and planar molecule consisting of extending methyl groups branching from a nitrogen molecule on each end of the carbon ring center [16]. TMCES is a bulkier molecule presenting itself in a tripod form with branches consisting of different atoms.

In examining these molecules it is desired to see the effect of the length of the molecule and the contribution of the methyl groups on electron transport when placed in the vicinity of the active site.

The charges throughout each of the mediators have been summarized in Tables II and III and the corresponding molecule optimized geometries shown in Figures 3 and 4 respectively. This charge distribution shows how each mediator could obtain and transfer an electron through their oppositely charged ends.

a)

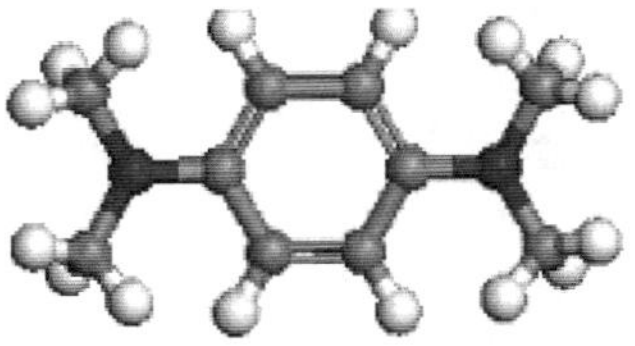

b)

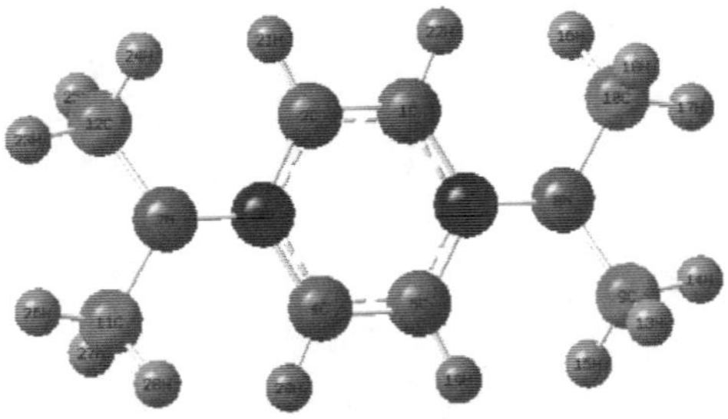

Figure 3. **a)** Depiction of TMPD molecule. **b)** Labeled atoms for assigned charges on TMPD molecule.

a)

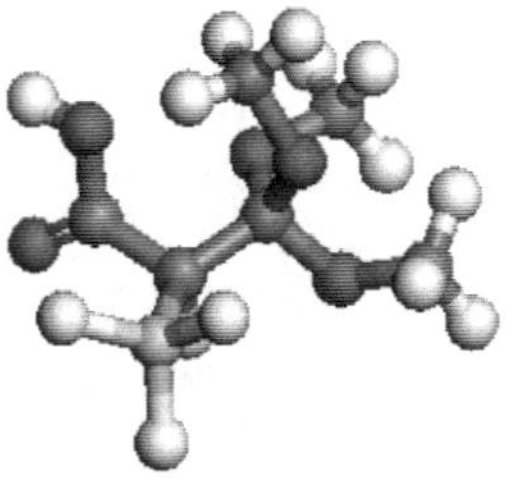

b)

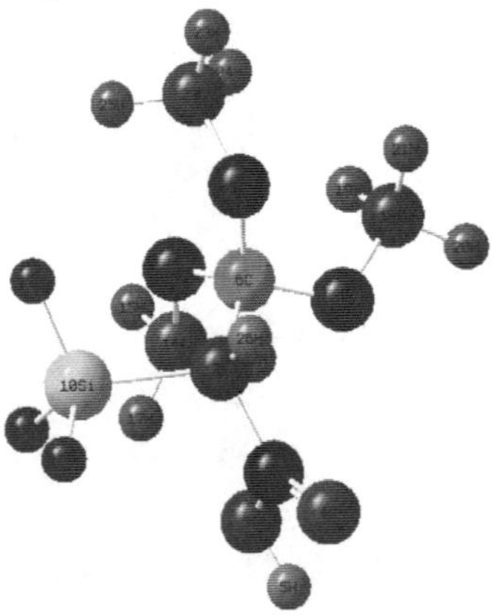

Figure 4. **a)** Depiction of TMCES molecule. **b)** Labeled atoms for assigned charges on TMCES molecule.

Atom Label	Composition	Charge	Atom Label	Composition	Charge
1	C	-0.21	15	H	0.16
2	C	-0.21	16	H	0.16
3	C	-0.03	17	H	0.15
4	C	-0.21	18	H	0.16
5	C	-0.21	19	H	0.12
6	C	-0.03	20	H	0.12
7	N	0.09	21	H	0.12
8	N	0.09	22	H	0.12
9	C	-0.40	23	H	0.15
10	C	-0.40	24	H	0.16
11	C	-0.40	25	H	0.16
12	C	-0.40	26	H	0.15
13	H	0.16	27	H	0.16
14	H	0.15	28	H	0.16

Table II. Merz-Singh-Kollman (MK) charges on atoms of TMPD molecule.

Atom Label	Composition	Charge	Atom Label	Composition	Charge
1	C	-0.03	14	C	-0.29
2	C	-0.16	15	H	0.17
3	O	-0.20	16	H	0.18
4	O	-0.26	17	H	0.19
5	H	0.28	18	C	-0.29
6	C	-1.04	19	H	0.18
7	O	0.02	20	H	0.17
8	O	0.01	21	H	0.18
9	O	-0.05	22	C	-0.22
10	Si	0.57	23	H	0.17
11	H	-0.10	24	H	0.17
12	H	-0.09	25	H	0.18
13	H	-0.09	26	H	0.27

Table III. Merz-Singh-Kollman (MK) charges on atoms of TMCES molecule.

Next the chosen mediator molecules were placed in the presence of PQQ and a calcium +2 ion (Figures 5-8). The distance between the mediator molecule and the calcium ion was calculated (Table IV) to better examine the ease at which the mediator was allowed to penetrate the active site. The distances between the two of the oxygen molecules and the nitrogen atom of the PQQ molecule to the calcium ion (see Figure 2) were measured to provide insight to conformational changes of the active site model.

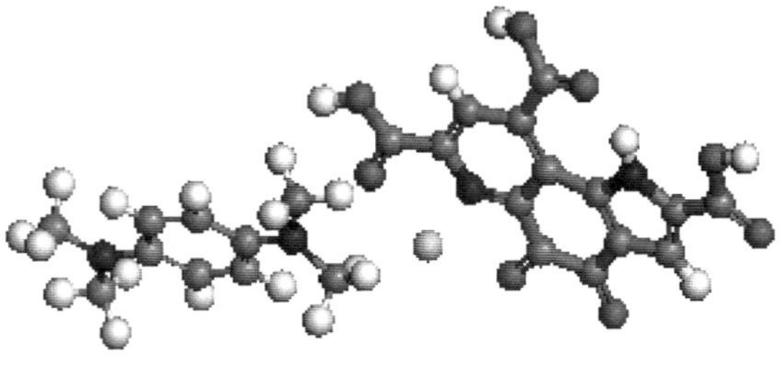

Figure 5. Geometry Optimization using GGA DFT for TMPD in the presence of PQQ and Ca^{2+}ion.

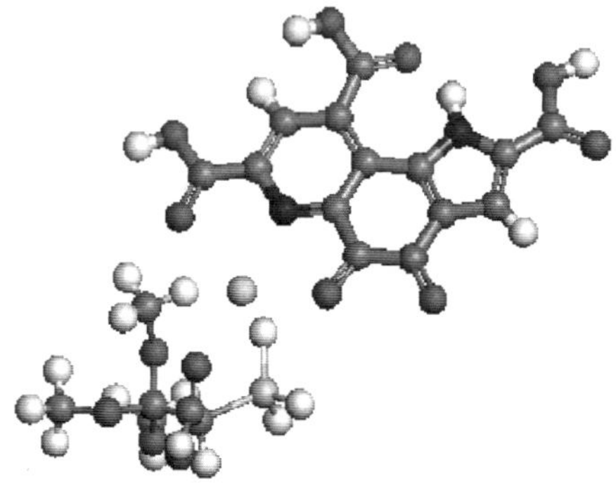

Figure 6. Geometry Optimization using GGA DFT for TMCES in the presence of PQQ and Ca^{2+}ion.

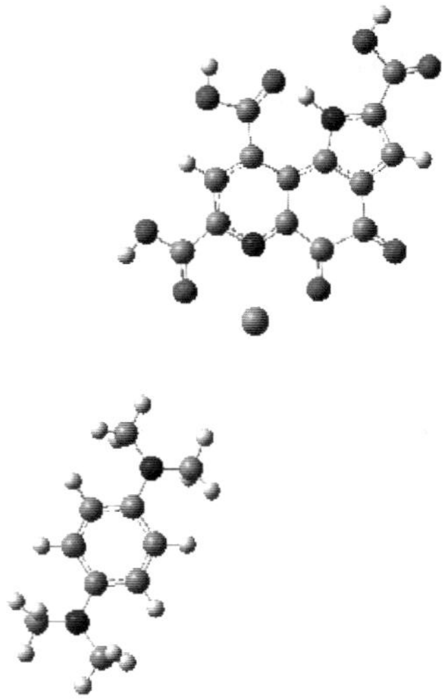

Figure 7. Geometry Optimization using hybrid DFT for TMPD in the presence of PQQ and Ca^{+2} ion.

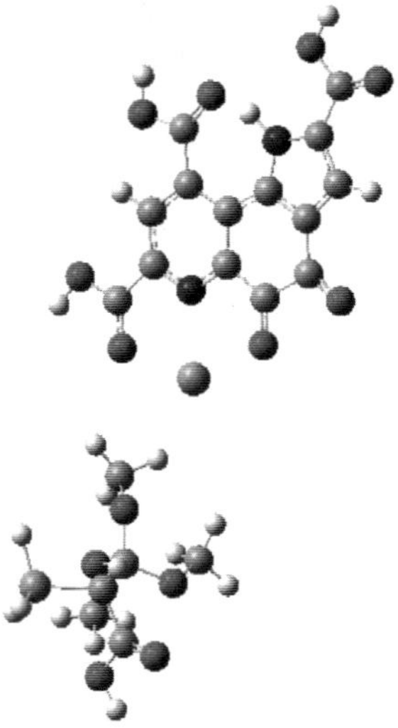

Figure 8. Geometry Optimization using hybrid DFT for TMCES in the presence of PQQ and Ca^{+2} ion.

	Mediator	Figure Number	Calcium to Nitrogen	Calcium to O1	Calcium to O4	Calcium to Mediator
DFT - GGA	TMPD	5	2.347	2.316	2.367	3.245
DFT - Hybrid	TMPD	6	2.475	2.470	2.529	4.191
DFT - GGA	TMCES	7	2.408	2.274	2.608	2.280
DFT - Hybrid	TMCES	8	2.398	2.439	2.424	2.400

Table IV. Distances (Angstroms) from PQQ, Ca^{2+}, and mediator systems referenced in Figures 5-8.

The two methods show that the calcium ion remains located near the nitrogen atom and between the two surrounding oxygen atoms. When the mediator molecule is added to the system the distance from the calcium ion to the nitrogen atom slightly lowers. TMCES more closely approaches the PQQ and Ca^{2+} system.

Finally, thirteen amino acids were added to the system to complete the active site center. In order to determine the most likely position of the mediator, minimization calculations were run to find the orientation of the mediator that provided the lowest energy configuration of the system. This procedure was followed for both TMPD and TMCES, as shown in Figure 9 and Figure 10, respectively. Once again the distances from the calcium ion to various atoms of PQQ and the mediator were measured. These same measurements were evaluated for all ten configurations of each mediator molecule. TMCES was closer to the calcium ion than TMPD. From all of the trials, the minimum energy conformation maximized the Calcium – Nitrogen distance for both TMCES and TMPD. As the size of the system increased the values obtained remained relatively consistent with those found experimentally using x-ray crystallography.

	TMPD	TMCES	Experimental[2]
Ca 2+ -N	3.749	2.568	2.4
Ca 2+ -O1	2.337	2.413	2.5
Ca 2+ -O4	4.388	2.885	2.44
Ca 2+ -mediator	2.459	2.75	X

Table V. Distances (Angstroms) for the minimum energy configuration of the complete active site with mediator present.

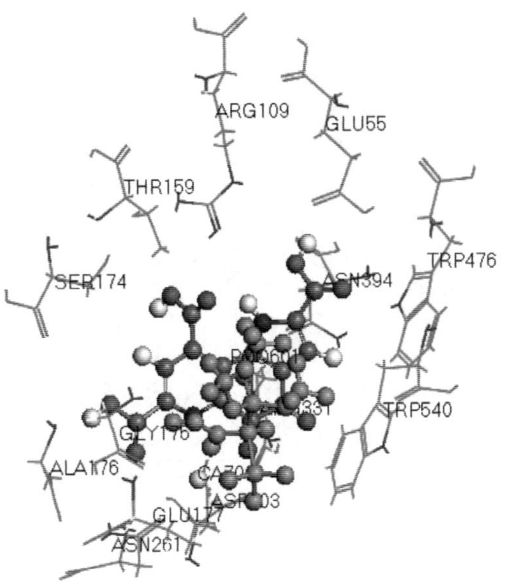

Figure 9. Minimum energy conformation for the optimized geometry of the Active Site and TMPD.

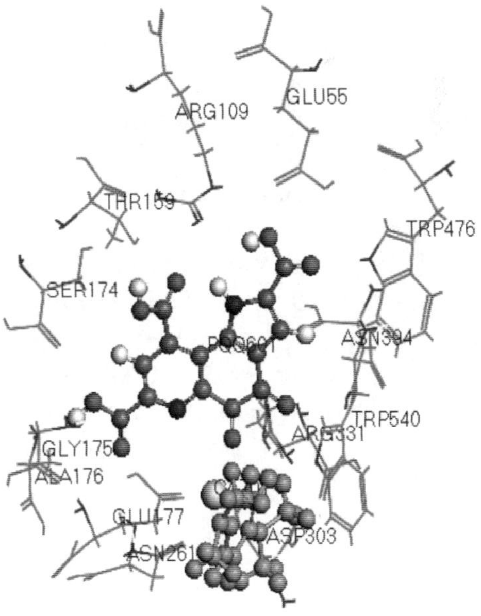

Figure 10. Minimum energy conformation for the optimized geometry of the Active Site and TMCES. A side view is shown to the right to clarify.

Ionization Potential of Mediators

The ionization energy of an atom or molecule can be defined as the energy required to remove one electron [17]. The ionization potential is then equal to the ionization energy divided by the charge of an electron. This is also sometimes described as the electron binding energy which is defined as the energy required to release an electron from its atomic or molecular orbital. In the case of our system, the ionization potential describes when a charged molecule loses the electron to become neutral. The positive values that are reported imply that energy is required to release the electron from the mediator molecule.

$$\text{I. P.} = [E\,(\text{molecule}^{+}) - E\,(\text{molecule}^{0})]\,/\,e^{-}\ \text{charge} \qquad (1)$$

The ionization potentials of TMPD and TMCES are 15.3 V and 18.7 V, respectively. This implies that more energy is required to release the electron from the TMCES molecule. Although this means that it is more difficult for TMCES to give up an electron, it also compliments its ability to retain the electron within the molecule without true concern of releasing the electron to surrounding amino acids which may disrupt the electron transfer to the substrate. The calculated ionization potentials show that the TMPD and TMCES molecules are in close comparison to the calculated ionization potential of PQQ, 21.7 V. Most importantly, this calculation shows that the transfer of an electron from TMCES requires more energy to release an electron and it has the desirable ability to retain the electron, similar to PQQ.

Electron Affinity of Mediators

The electron affinity of an atom or molecule can be defined as the energy released when an electron is attached to a neutral atom or molecule [18]. When this term is applied to our system it is describing when an electron has moved to the previously neutral mediator molecule [19]. A negative value shows the energy is released when the electron is obtained thereby stabilizing the system.

$$\text{E.A.} = E\,(\text{molecule}^{0}) - E\,(\text{molecule}^{-}) \qquad (2)$$

The electron affinities of TMPD and TMCES are -0.92 eV and -0.39 eV, respectively. When the neutral molecule obtains an electron energy is released into the system. This implies that more energy is released when an electron is obtained by TMPD. The calculated electron affinity of PQQ is 2.28 eV, which supports the investigation to a more receptive molecule. TMCES has a smaller electron affinity which implies that less energy will be released to the overall system once the electron is obtained. This is important in maintaining a stable system.

Conclusions

The theory level of the computational methods utilized in this work was first verified by comparing lengths found computationally to those found experimentally. The charge distribution of the TMPD and TMCES were investigated to relatively show that each molecule has both positively and negatively charged ends.

The mediators were then placed in the presence of PQQ and Ca^{2+} to show that the distances between the calcium ion and certain atoms of the PQQ molecule remain close but decrease in the presence of the mediators. Although the structure of TMPD seems to have the attributes to move the electron through the mediator molecule, it does not show promising results in penetrating into the active site to initially retrieve the electron from the active center of the enzyme. The branches of the TMCES molecule do allow for the transport of the electron through the mediator and although the molecule is bulky it still penetrates into the active site

The mediators were then individually placed in ten different orientations within the complete active center. Geometry Optimizations were performed and the minimum energy configurations were shown with the previous distances evaluated again. Within the ten trials themselves, the minimum energy configuration gave the structural maximum Calcium-Nitrogen bond for both TMCES and TMPD.

Lastly, the ionization potential and electron affinity calculations provided insight to the effectiveness of the mediator molecule to both retrieve and release electrons in an effort to facilitate electron transfer to the substrate. TMCES had the more desirable ionization potential of 18.3 V, showing it has the ability to release an electron, but with the control to maintain a relatively stable system. TMCES also had the smaller, more desirable electron affinity. This calculation supported the ability of TMCES to release energy and stabilize the system upon obtaining an electron. For all of the geometric and energetic parameters described, it has been elucidated that TMCES may be a more appropriate mediator to better facilitate electron transfer.

Acknowledgements

It is gratefully acknowledge the support of the National Science Foundation NSF-CAREER grant CTS-0449046 and grant DMR-0414903 that funded the SGI Altix3700 computational resource used for this study.

References

1. Ghosh M., Anthony C., Harlas K., Goodwin M. G., and Blake C. C. F., *The Refined Structure of the Quinoprotein Methanol Dehydrogenase from Methylobacterium Extorquens at 1.94 Å.* Structure (London), 1995. 3: p. 1771-1787.
2. Xia Z. X., He Y. N., Dai W. W., White S., Boyd G., and Mathews F. S., *Detailed Active Site Configuration of a New Crystal Form of Methanol Dehydrogenase from Methylophilus W3a1 at 1.9 Å Resolution.* Biochem. J., 1999. 38: p. 1214-1220.
3. Williams P. A., Coates L., Mohammed F., Gill R., Erskine P. T., Coker A., Wood S. P., Anthony C. and Cooper B., *The atomic resolution structure of methanol dehydrogenase from Methylobacterium extorquens.* Acta Crystallographica Section D, 2005. D61: p. 75-79.
4. Xia Z. X., Dai W. W., Xiong J. P., Hao Z. P., Davidson V. L., White S., and Mathews F. S., *The Three-Dimensional Structures of Methanol Dehydrogenase from Two Methylotrophic Bacteria at 2.6 Å Resolution.* J. Biol. Chem., 1992. 267: p. 22289-22297.
5. Anthony C., *Methanol Dehydrogenase, a PQQ-Containing Quinoprotein Dehydrogenase*. Subcellular Biochemistry, ed. E. Holzenburg and Scrutton. 2000, New York.: Kluwer Academic/ Plenum Publishers. 73-118.
6. White S., Boyd G., Mathews F. S., Xia Z. X., Dai W. W., Zhang Y. S., and Davidson V. L., *The Active Site Structure of Calcium Containing Methanol Dehydrogenase.* Biochemistry, 1993. 32: p. 12955-12958.

7. H.M. Berman, J. Westbrook, Z. Feng, G. Gilliland, T.N. Bhat, H. Weissig, I.N. Shindyalov, P.E. Bourne: The Protein Data Bank. Nucleic Acids Research, 28 pp. 235-242 (2000).
8. E. Katz, I. Willner and B. Willner, "Electrical Contact of Redox Enzyme Layers Associated with Electrodes: Routes to Amperometric Biosensors." *Electroanalysis*, 1997. 9(13): p. 965 - 977.
9. Williams, P., Coates, L., Mohammed, F., Gill, R., Erskine, P., Bourgeois, D., Wood, S.P., Anthony, C., Cooper, J.B. The 1.6A X-Ray Structure of the Unusual C-Type Cytochrome, Cytochrome cL, from the Methylotrophic Bacterium Methylobacterium Extorquens. J.Mol.Biol. v357 pp. 151-162, 2006
10. A. Ranta, A. Halme and X. C. Zhang, "Effect of Different Catalytic Oxidants on the Performance of a Biocatalytic Methanol Fuel Cell." *Proceedings 204th Meeting of The Electrochemical Society*, 2003.
11. X. C. Zhang, A.Ranta, and A. Halme, "Direct Methanol Biocatalytic Fuel Cell-Considerations of Restraints on Electron Transfer." *Biosensors and Bioelectronics*, 2006. 21: p. 2052-2057.
12. A. Halme, X. Zhang, A. Ranta, *"Study of Biological Fuel Cells,"* in Small Fuel Cells 2000, New Orleans, 26-28 April 2000.
13. R.N. Patel, A. Felix, "Microbial Oxidation of Methane and Methanol: Crystallization and Properties of Methanol Dehydrogenase from *Methylosinus sporium*." Journal of Bacteriology, 1976, 128(1): p. 413-424.
14. Gaussian 03, Revision C.02, M. J. Frisch, G. W. Trucks, H. B. Schlegel, G. E. Scuseria, M. A. Robb, J. R. Cheeseman, J. A. Montgomery, Jr., T. Vreven, K. N. Kudin, J. C. Burant, J. M. Millam, S. S. Iyengar, J. Tomasi, V. Barone, B. Mennucci, M. Cossi, G. Scalmani, N. Rega, G. A. Petersson, H. Nakatsuji, M. Hada, M. Ehara, K. Toyota, R. Fukuda, J. Hasegawa, M. Ishida, T. Nakajima, Y. Honda, O. Kitao, H. Nakai, M. Klene, X. Li, J. E. Knox, H. P. Hratchian, J. B. Cross, V. Bakken, C. Adamo, J. Jaramillo, R. Gomperts, R. E. Stratmann, O. Yazyev, A. J. Austin, R. Cammi, C. Pomelli, J. W. Ochterski, P. Y. Ayala, K. Morokuma, G. A. Voth, P. Salvador, J. J. Dannenberg, V. G. Zakrzewski, S. Dapprich, A. D. Daniels, M. C. Strain, O. Farkas, D. K. Malick, A. D. Rabuck, K. Raghavachari, J. B. Foresman, J. V. Ortiz, Q. Cui, A. G. Baboul, S. Clifford, J. Cioslowski, B. B. Stefanov, G. Liu, A. Liashenko, P. Piskorz, I. Komaromi, R. L. Martin, D. J. Fox, T. Keith, M. A. Al-Laham, C. Y. Peng, A. Nanayakkara, M. Challacombe, P. M. W. Gill, B. Johnson, W. Chen, M. W. Wong, C. Gonzalez, and J. A. Pople, Gaussian, Inc., Wallingford CT, 2004.
15. Jensen F., *Introduction to Computational Chemistry*. 2002: Wiley&Sons.
16. L. Boilet, G. Buntinx, C. Lefumeux, and O. Poizat,"Ultrafast photoinduced electron transfer from N,N,N',N'-tetramethyl-p-phenylenediamine and N,N,N',N'-tetramethylbenzidine to dichloromethane." Journal of Photochemistry and Photobiology A: Chemistry 163 (2004) p. 529-536.
17. M. Vijayan, CRC Handbook of Biochemistry and Molecular Biology. 3rd ed. Proteins, ed. G.D. Fasman. Vol. 2. 1976, Cleveland: *CRC Press*. 742- 759.
18. Handbook of Chemistry and Physics. 69 ed. 1988, Boca Raton, Florida: *CRC Press*, Inc.
19. Gokel, George W. Dean's Handbook of Organic Chemistry (2nd Edition). (pp. 11). McGraw-Hill.

Methanol Electro-Oxidation By Ion-Modified Methanol Dehydrogenase Enzymes

Nagesh B. Idupulapati and Daniela S. Mainardi
Louisiana Tech University, Ruston, LA

Abstract

The catalytic oxidation of methanol to formaldehyde by MDH is investigated theoretically at the B3LYP/6-31++g** theory level by considering a small active site model (PQQ, Ca^{2+}, ASP) in the presence of methanol. Both proposed reaction mechanisms, the Addition-Elimination and Hydride-Transfer mechanisms are investigated. The rate-determining step in both cases is found to be the first one which involves proton transfer from methanol to a base catalyst (ASP) and cleavage of C-H bond of methanol from the energy barrier calculations.

1. Introduction

Methanol Dehydrogenase (MDH) is a water soluble enzyme able to oxidize methanol, a relatively cheap fuel and other primary alcohols to their corresponding aldehydes [1].The crystal structure of MDH from *Methlylobacterium extorquens* [2-4] and from *Methylophilus W3A1*[5, 6] has been characterized using X-ray crystallographic methods. From all these studies, it has been determined that the MDH active site contains Pyrroloquinoline Quinone (PQQ) as its co-factor and Ca^{2+} ion (Figure 1). Apart from holding the PQQ molecule in place in the active site, the Ca^{2+} is believed to have an important role in the electro-oxidation reaction mechanism [1].

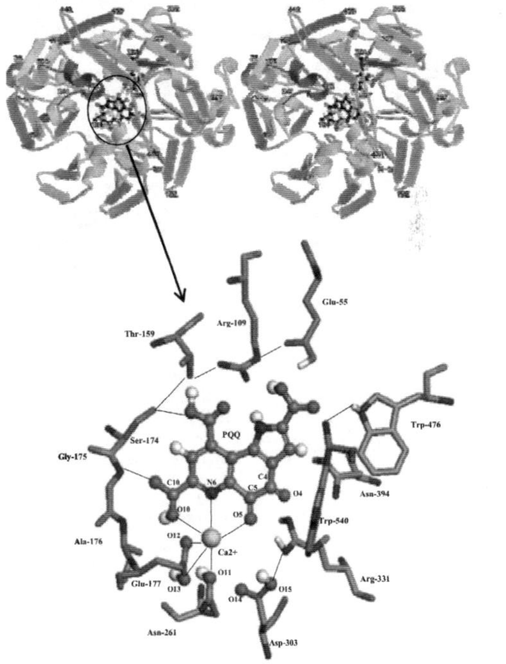

Figure 1. Methanol Dehydrogenase Enzyme from *Methlylobacterium extorquens* and its corresponding active site (magnified).

Two possible mechanisms for methanol oxidation by MDH have been proposed [1]. The first one is the Addition-Elimination mechanism (Figure 2a) that involves a proton transfer from the alcohol group to an active site base, which is proposed to be Asp-303. The presence of this catalytic base at the active site initiates the reactions by abstracting a proton from the substrate (methanol), which is followed by the formation of covalent hemiketal intermediate since the oxyanion (O16⁻) produced by the proton abstraction attacks the electrophilic C5 of PQQ (Figure 2a). The final step is characterized by the proton transfer to O5 and then the methyl proton H17 is abstracted by O4 of PQQ resulting in the formation of the by-product formaldehyde (CH_2O) [1, 7]. This mechanism is supported by Itoh et al. [8] by their theoretical calculations and experiments on PQQ systems in organic solution. These authors suggested that in the case of the Ca^{2+} complex of the C5 hemiketal intermediate, the metal ion is coordinated by both oxygen atoms (O5 of PQQ and methyl oxygen O10) and this coordination forces the methyl group to flip towards the O4 carbonyl oxygen, suggesting that this O4 might act as a base for methyl proton abstraction of H17 [8]. The second proposed mechanism is the Hydride transfer mechanism (H-T) (Figure 2b), in which the initial proton abstraction is the same as step 1 of the Addition-Elimination (Figure 2a), but in this case the electrophilic C5 of PQQ is involved directly in the removal of methyl hydrogen of methanol resulting in the formation of the byproduct formaldehyde [1, 7]. This mechanism is supported by various studies [2, 6, 9-11] of which a recent study of MDH represented a structure where the PQQ is in the reduced form with C5 carrying a hydroxyl group [6]. A similar structure was also inferred from the x-ray structure of mutant MDH in which Asp-303 was changed by site-directed mutagenesis to Glu-303. This glutamate was able to form a hydrogen bond with O5 oxygen of PQQ which would be possible if there is hydroxyl in this position. Both of these studies lend considerable support to the hydride transfer mechanism of MDH [2].

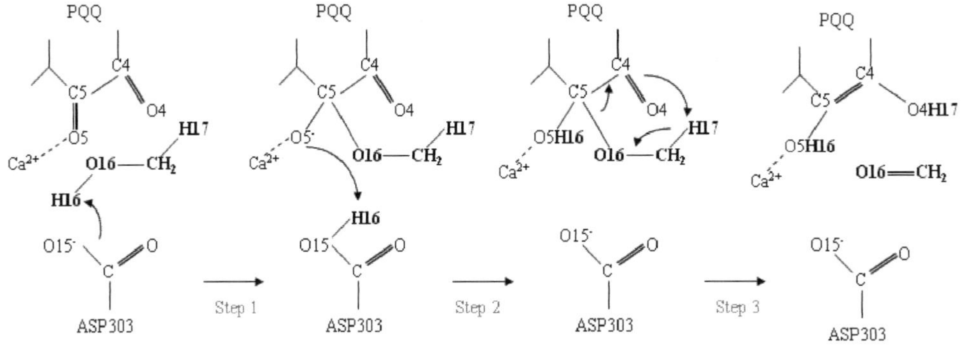

(a)

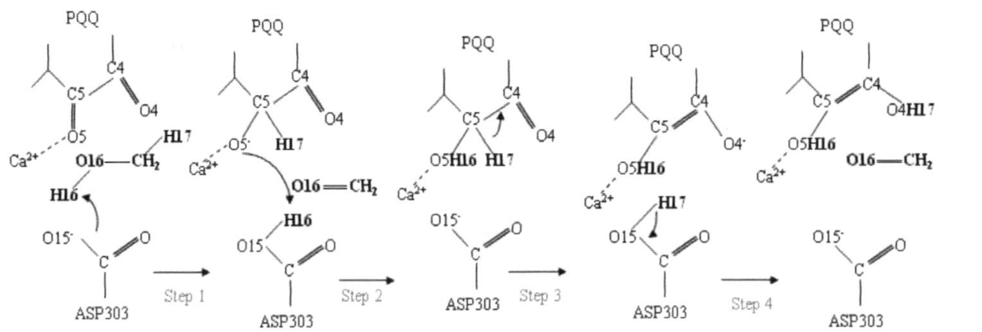

(b)

Figure 2. (a) Addition-elimination (A-E) mechanism proposed for MDH **(b)** Hydride Transfer mechanism proposed for MDH.

These studies suggest that the oxidation of methanol by MDH supports the hydride transfer mechanism, and consists of the association of methanol to active site Ca^{2+} followed by de-protonation of Ca^{2+}-bound methanol by Asp-303, hydride transfer from Ca^{2+}-bound substrate to the carbonyl at C4 position (Figure 2b), formation of a Ca^{2+}- bound formaldehyde, and finally the release of formaldehyde from the active site of the enzyme [11]. More studies [9] using Molecular Dynamics (MD) simulations suggested that Asp303-CO_2^- is the general base catalyst for removing the hydroxyl proton of methanol and transfer it to the C5 carbonyl of PQQ (Figure 2b). No consensus appears in the literature about the base catalyst, when practically at the same time, the same research group[10], using MD simulations to study the PQQ reduction in aqueous solution suggested that Glu-177 is the general base catalyst rather than the previously suggested Asp-303.

Recently Leopoldini et al. [12] investigated the two proposed mechanisms at the density functional B3LYP theory level. The employed active site model included PQQ cofactor, Ca^{2+} and some nearby residues (relevant portions of the residues close to PQQ were considered and hydrogen atoms of those were frozen to prevent unreliable expansion of model during optimization procedure). Their calculated energy barriers for the rate-determining step (34.6 kcal/mol for A-E, 32.3 kcal/mol for H-T) suggested that the two proposed mechanisms would be incompatible to the general kinetic requirements of an enzymatic process (15-20 kcal/mol). A third mechanism (Addition-elimination-protonation) was proposed by tem, in which the sequence of the steps involves the cleavage of the C-H bond in the substrate before the cofactor protonation by the Asp amino acid residue was found to be more reliable as the activation energy in the rate-determining step is 16 kcal/mol [12].

To the best of our knowledge, no perfect agreement has been reached by different researchers from their experimental and theoretical studies about which mechanism actually proceeds for the oxidation of methanol in the active site of MDH. In this work, a detailed theoretical investigation is carried out on MDH active site model cluster in the presence of methanol. The model considers PQQ, Ca^{2+}, and the residue ASP since this amino acid is the best supported as the base catalyst for proton transfer. No constrains (frozen atoms) are used during the optimization procedures and the energy barriers are obtained for each step including the rate-determining step for addition elimination and hydride transfer mechanisms (Figure 2a and b).

2. Computational Details

Applied quantum chemical methods such as DFT have been successfully used for getting information about geometry and energetics of atomic and molecular systems, activation energies, reaction paths and mechanisms. All the computations reported in this work are carried out with Gaussian 03 program (revision C.02) [13] employing the hybrid Becke exchange and Lee, Yang and Parr correlation (B3LYP) functional and 6-31++g** basis set. Frequency calculations are performed at same level of theory on all stationary points of reaction paths with the aim of evaluating their character of minima or saddle points and also to obtain the zero-point corrected energies for calculating the reaction barrier values. Intrinsic Reaction Coordinate (IRC) calculations are performed to find out whether the obtained transition state connects the reactant and product.

3. Results and Discussion

Both Addition-Elimination (A-E) and Hydride Transfer (H-T) mechanisms are investigated at the B3LYP/6-31++G** theory level. The active site model (A) considered for evaluating both oxidation mechanisms consisted of PQQ, Ca^{2+}, ASP, and its chemistry tested upon the presence of methanol (M). Transition states (TS) and intermediates (INT) obtained from both mechanisms are presented and discussed. Later gas-phase potential energy surfaces of both mechanisms are analyzed and discussed.

3.1 Addition-Elimination Mechanism (A-E)

The calculated first step in the A-E mechanism (Figure 3) shows that the distance C5-O16 from the initial reactant (A + M), TS1, and INT1 is reduced from 4.38 to 2.46 and 1.41 Å respectively, suggesting the formation of hemiketal intermediate of PQQ with methanol. It is also observed that H16 from methanol is attached to O14 of the aspartic acid suggesting the abstraction of proton from alcohol by the amino acid base (ASP). Also the distance between 04 of PQQ and H17 of methanol is decreased from 5.33 (initial reactant) to 2.81 Å (INT1) which implies that the formation of formaldehyde by-product could be possible.

The second step shows that O5-H16 decreases from 1.43 (INT1) to 1.22 (TS2) to 0.97 (INT2) Å indicating that there is a proton transfer from O14 of ASP to C5 of PQQ. Also the C5-O16 distance increases slightly from INT1 to INT2 with the hemiketal orienting in a way such that the other proton transfer (H17 to O4 of PQQ) would be more possible. The final step which involves INT2, TS3 and AH2 + CH2O shows that C5-O16 distance increases gradually from 1.48 to 3.08 to 3.53 Å and O4-H17 distance decreases from 2.70 to 1.03 to 0.97 Å respectively indicating that by-product formaldehyde is formed by the second proton transfer.

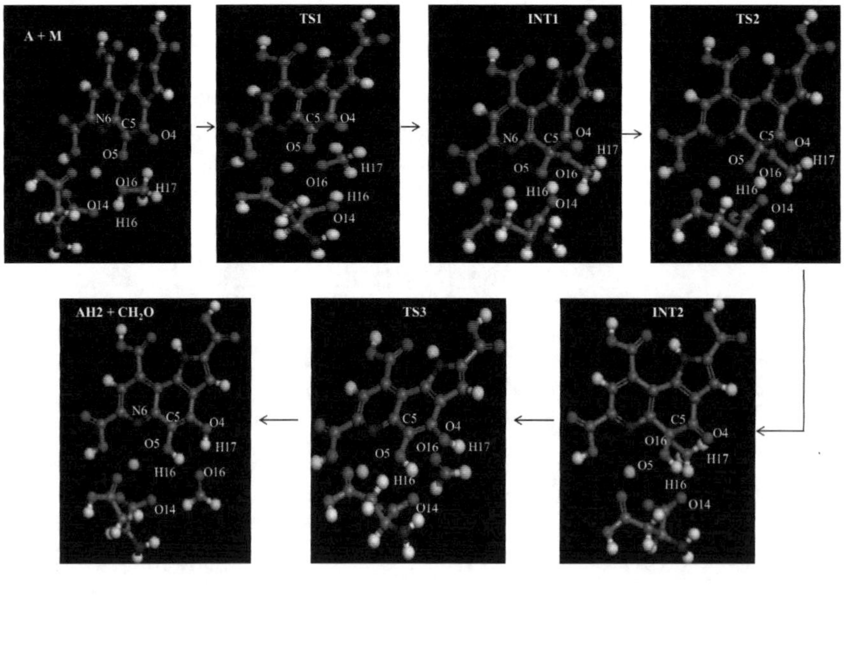

Figure 3. Ground State Geometries of Intermediates, Transition states involved in Addition-Elimination Mechanism (A+M: PQQ, Ca^{2+}, ASP, Methanol ; TS: Transition state ; INT: Intermediates ; AH2 + CH2O: PQQH2, Ca^{2+}, ASP, Formaldehyde).

3.2 Hydride Transfer Mechanism (H-T)

The calculated first step in the H-T mechanism (Figure 4) involves A + M, TS1 and INT1. According to our calculations, the distance O14-H16 is reduced from 2.18 to 0.99 to 0.97 Å and the distance C5-H17 is reduced from 4.85 to 1.83 to 1.21 Å, indicating that the two hydride transfers (H16, H17) would be possible.

The second step shows that the distance O14-H16 increased from 1.08 to 1.28 to 3.03 Å with respect to INT1, TS2 and INT2. The C5-H17 distance remains almost the same. This increase in distance shows that there would be a proton transfer from O14 of ASP to O5 of PQQ.

The third step involves INT2, TS3 and INT3. In this step, the distance C5-H17 has been increased from 1.11 to 4.36 to 6.30 Å and O14-H17 decreased from 5.22 to 2.62 to 1.01 Å respectively, indicating that hydride transfer from C5 of PQQ to O14 of ASP would be possible.

The final step involves INT3, TS4 and AH2 + CH2O. In this case, the distance O14-H17 has been increased from 1.01 to 2.36 to 6.38 Å and O4-H17 decreased from 6.27 to 4.26 to 1.01 Å respectively, indicating that PQQ is finally reduced.

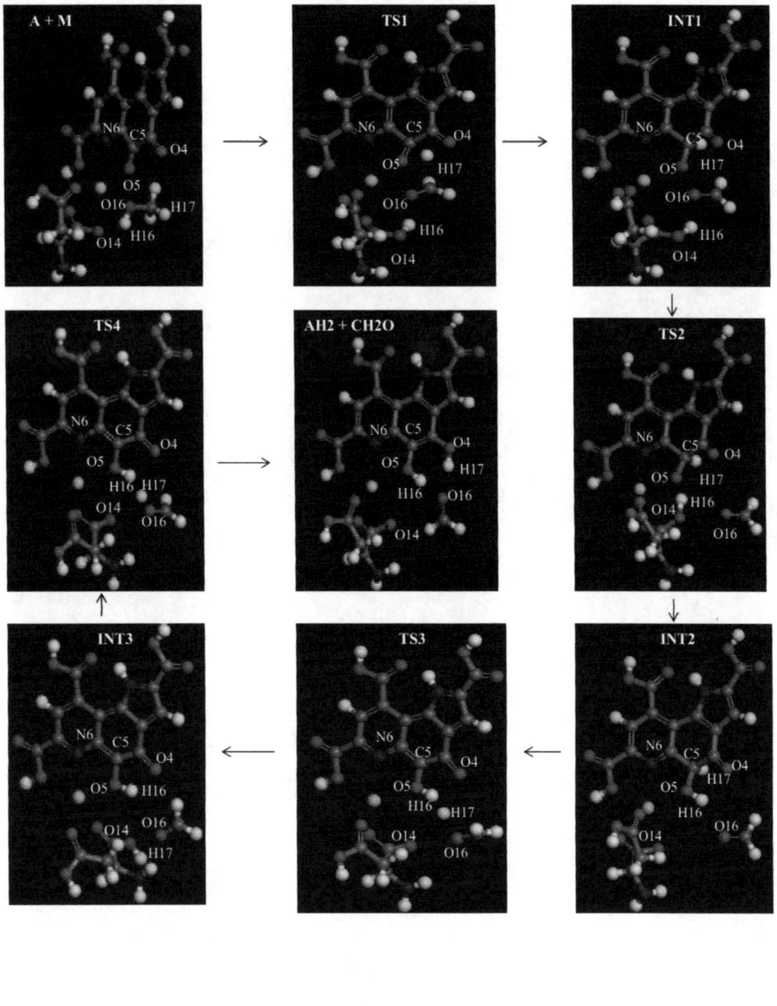

Figure 4. Ground State Geometries of Intermediates, Transition states involved in Hydride-Transfer Mechanism (A+M: PQQ, Ca^{2+}, ASP, Methanol; TS: Transition state; INT: Intermediates; AH2 + CH2O: PQQH2, Ca^{2+}, ASP, Formaldehyde).

3.3 Gas-Phase Potential Energy Surface for A-E and H-T Mechanisms

The zero-point corrected ground state energies of the initial reactant, intermediates, and transition states obtained are plotted for the A-E mechanism and H-T mechanisms (Figure 5a and b). Reaction barriers are calculated for each step under investigation. The reaction barriers energies listed in the Figure 5a for step 1 of A-E mechanism suggest that the forward reaction barrier (20.4 Kcal/mol) is greater than the reverse reaction barrier (4.4 Kcal/mol), indicating that the reverse reaction is more favorable than the forward reaction. For steps 2 and 3 of A-E mechanism (Figure 5a), the reverse reaction barrier is higher than the forward barrier, implying that the forward reactions in both cases are more favorable.

Same is the case with step 1 of H-T mechanism (Figure 5b) where the forward reaction barrier value is higher than the reverse reaction barrier value. Steps 2, 3 & 4 are much more favorable in the forward direction because of their lower energy barriers.

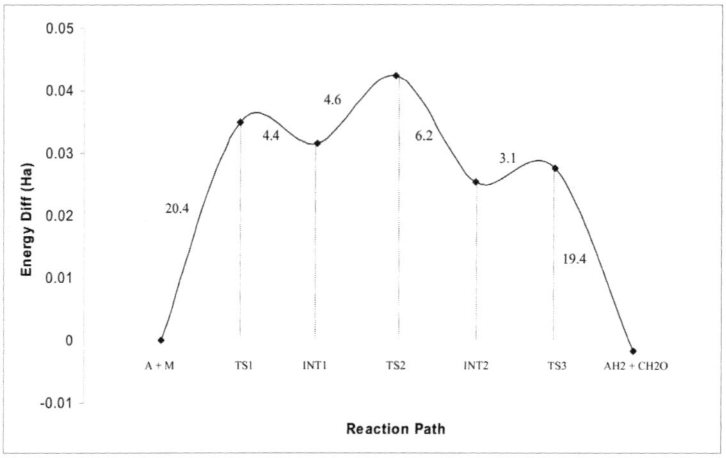

(a)

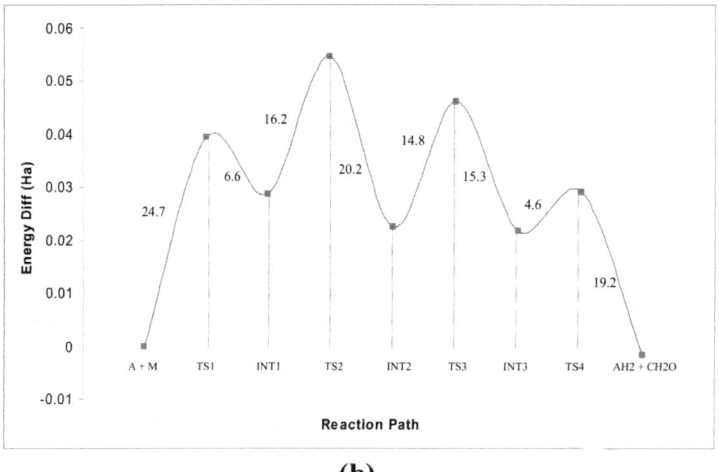

(b)

Figure 5. Gas-Phase Potential Energy Surface for **(a)** A-E and **(b)** H-T. Energy barriers calculated are in Kcal/mol cases.

4. Conclusions

The two oxidation mechanisms were investigated in gas-phase at the DFT B3LYP/6-31++g** level by employing important portions of active site of MDH in the presence of methanol considering no constraints in the calculations. The first step in both proposed Addition-Elimination and Hydride Transfer mechanisms is not favorable at this theory level since the forward reaction barrier values was higher than the reverse barrier. Better theory levels are being evaluated and the addition of solvent effects in these calculations considered in order to check the reliability of the presented results.

5. Acknowledgements

We would like to thank the financial support of National Science Foundation NSF-CAREER grant CTS-0449046 and also DMR-0414903 that funded the SGI Altix3700 computational resource used for this study.

References

1. Anthony C., *Methanol Dehydrogenase, a PQQ-Containing Quinoprotein Dehydrogenase*. Subcellular Biochemistry, ed. E. Holzenburg and Scrutton. 2000, New York.: Kluwer Academic/ Plenum Publishers. 73-118.
2. Afolabi P. R., F. M., Amaratunga K., Majekodunmi O., Dales S. L., Gill R., Thompson D., Cooper J. B., Wood S. P., Goodwin P. M., and Anthony C., *Site-Directed Mutagenesis and XRay-Crystallography of the PQQ-Containing Quinoprotein Methanol Dehydrogenase and Its Electron Acceptor, Cytochrome CL*. Biochem. J., 2001. **40**: p. 9799-9809.
3. Ghosh M., Anthony C., Harlas K., Goodwin M. G., and Blake C. C. F., *The Refined Structure of the Quinoprotein Methanol Dehydrogenase from Methylobacterium Extorquens at 1.94 Å*. Structure (London), 1995. **3**: p. 1771-1787.
4. White S., Boyd G., Mathews F. S., Xia Z. X., Dai W. W., Zhang Y. S., and Davidson V. L., *The Active Site Structure of Calcium Containing Methanol Dehydrogenase.* Biochemistry, 1993. **32**: p. 12955-12958.
5. Xia Z. X., Dai W. W., Xiong J. P., Hao Z. P., Davidson V. L., White S., and Mathews F. S., *The Three-Dimensional Structures of Methanol Dehydrogenase from Two Methylotrophic Bacteria at 2.6 Å Resolution.* J. Biol. Chem., 1992. **267**: p. 22289-22297.
6. Xia Z. X., He Y. N., Dai W. W., White S., Boyd G., and Mathews F. S., *Detailed Active Site Configuration of a New Crystal Form of Methanol Dehydrogenase from Methylophilus W3a1 at 1.9 Å Resolution.* Biochem. J., 1999. **38**: p. 1214-1220.

7. Anthony C. and Williams P., *The Structure and Mechanism of Methanol Dehydrogenase.* Biochim Biophys Acta, 2003. **1647**: p. 18-23.
8. Itoh S., Kawakami H. and Fukuzumi S., *Model Studies on Calcium-Containing Quinoprotein Alcohol Dehydrogenases. Catalytic Role of Ca2+ for the Oxidation of Alcohols by Coenzyme PQQ (4,5 Dihydro-4,5-Dioxo-1h-Pyroolo[2,3-F]Quinoline-2,7,9-Tricarboxylic Acid).* Biochemistry, 1998. **37**(6562-6571).
9. Reddy S. Y., Mathews F. S., Zheng Y. J. and Bruice T. C., *Quinoprotein Methanol Dehydrogenase: A Molecular Dynamics Study and Comparison with Crystal Structure.* J. Mol.Struc., 2003. **655**: p. 269-277.
10. Reddy S. Y. and Bruice T. C., *In-Silico Studies of the Mechanism of Methanol Oxidation by Quinoprotein Methanol Dehydrogenase.* J. Am. Chem. Soc, 2003. **125**(8141-8150).
11. Zheng Y. J. and Bruice T. C., *"Conformation of Coenzyme Pyrroloquinoline Quinone and the Role of Ca2+ in the Catalytic Mechanism of Quinoprotein Methanol Dehydrogenase.* Proc. Natl.Acad. Sci., 1996. **94**(11881-11886).
12. Leopoldini M, Russo N, and Toscano M, *The Preferred Reaction Path for the Oxidation of Methanol by PQQContaining Methanol Dehydrogenase: Addition–Elimination versus Hydride-Transfer Mechanism.* Chemical European journal, 2007. **13**: p. 2109-2117.
13. H. B. S. G. W. T. M. J. Frisch, G. E. Scuseria, M. A. Robb, J. R.Cheeseman, J. A. Montgomery, Jr., T. Vreven, K. N.Kudin, J. C. Burant, J. M. Millam, S. S. Iyengar, J. Tomasi, V. Barone, B. Mennucci, M. Cossi, G. Scalmani, N. Rega, G. A. Petersson, H. Nakatsuji, M. Hada, M. Ehara, K. Toyota, R. Fukuda, J. Hasegawa, M. Ishida, T. Nakajima, Y. Honda, O. Kitao, H. Nakai, M. Klene, X. Li, J. E. Knox, H. P. Hratchian, J. B. Cross, C. Adamo, J. Jaramillo, R. Gomperts, R. E. Stratmann, O. Yazyev, A. J. Austin, R. Cammi, C. Pomelli, J. W. Ochterski, P. Y. Ayala, K. Morokuma, G. A. Voth, P. Salvador, J. J. Dannenberg, V. G. Zakrzewski, S. Dapprich, A. D. Daniels, M. C. Strain, O. Farkas, D. K. Malick, A. D. Rabuck, K. Raghavachari, J. B. Foresman, J. V.Ortiz, Q. Cui, A. G. Baboul, S. Clifford, J. Cioslowski, B. B. Stefanov, G. Liu, A.Liashenko, P. Piskorz, I. Komaromi, R. L. Martin, D. J. Fox, T. Keith, M. A. AlLaham, C. Y. Peng, A. Nanayakkara, M. Challacombe, P. M. W. Gill, B. Johnson, W.Chen, M. W. Wong, C. Gonzalez, and J. A. Pople, Gaussian 03, Revision C.02, 2003.

487d A First Principles Study Of Direct Electrooxidation Of Aqueous Borohydride On Au And Pt Surfaces

Michael Janik and Gholamreza Rostamikia. Chemical Engineering, Pennsylvania State University, 158 Fenske Laboratory, University Park, PA 16802

Direct borohydride fuel cells (DBFCs) offer the potential for direct chemical to electrical energy conversion from a high specific energy, water soluble fuel. Power densities similar to those attained from direct methanol fuel cells have been demonstrated [1], despite substantially less research focus. However, the lack of effective anode materials for the electrocatalysis of borohydride oxidation is the major limitation in advancing the application of these devices. This paper discusses the application of electronic structure modeling to determine the electrocatalytic reaction mechanism and design metal alloy catalysts for increasing the power density and efficiency of DBFCs. The development of optimal electrocatalysts for these processes could enable the use of aqueous borohydride solutions as a chemical energy storage vehicle for portable power applications. This presentation will discuss the use of density functional theory based electronic structure calculations to elucidate the elementary steps of borohydride oxidation at the metal-solution, electrochemical interface. Auxiliary benefits of mechanism determination are the elucidation of catalytic paths for hydrogen evolution from aqueous borohydride solutions and for reduction of the borate ion to produce/regenerate the borohydride fuel. The use of aqueous borohydride solutions (NaBH4) as an anode fuel in an alkaline fuel cell was demonstrated in the 1960's, however, the work of Amendola et al. in 1999 re-stimulated research in the area [2]. The ideal anode will oxidize BH_4^- (with $8OH^-$ delivered from the alkaline oxygen reduction cathode) to BO_2^- at low overpotentials producing 8 electrons per BH_4^- reactant molecule. However, high overpotentials and reduced Coulombic efficiency is observed due to both slow kinetics and non-selective hydrolysis reactions. We will present a mechanistic investigation of BH_4^- oxidation over the Au(111) surface, identifying stable surface species and rate limiting steps to motivate improved catalyst design. Periodic density functional theory calculations are used to determine elementary reaction energies and activation barriers. Multiple approaches to including the electrode potential dependence of both ion adsorption and surface reaction will be discussed, employing both metal-vapor and metal-solution model systems. Initial reaction steps will be compared over Pt(111) to determine the mechanism of non-selective hydrogen evolution, and determine the competing pathways which must be considered in electrocatalyst design.

References [1] C. Ponce de Leon, F. C. Walsh, D. Pletcher, D. J. Browning, J. B. Lakeman, J. Power Sources 155 (2006) 172. [2] S. C. Amendola, P. Onnerud, M. T. Kelly, P. J. Petillo, S. L. Sharp-Goldman, M. Binder, J. Power Sources 84 (1999) 130.

564a Modeling and Reduction of Carbon Dioxide Gas Formation at the Anode of a Direct Methanol Fuel Cell Using Chemically-Enhanced Solubility

Michael D. Lundin and Mark J. McCready. Chemical & Biomolecular Engineering, University of Notre Dame, 182 Fitzpatrick Hall, Notre Dame, IN 46556

The production of CO_2 gas at the DMFC anode leads to dramatic increases in pumping power requirements and reduced power output because of mass transfer limitations as bubble trains form in the channels of larger stacks. Experimental observations taken in a 5 cm^2 DMFC test cell operated at 60°C, 1 atm, and with a methanol/water fuel flow rates of 5-10 cc min^{-1} indicate that the rate of bubble formation can be reduced by increasing the fuel flow because more liquid is available for the CO_2 to dissolve in. Further observations indicate that KOH and LiOH added to the fuel eliminates CO_2 gas formation in situ at low concentrations because of the greatly increased solubility that results.

A mathematical model for the volumetric rate of CO_2 gas production that includes effects of temperature and solubility is developed and extended to include the effects of hydroxide ions in solution. The model is used to predict the onset location of gas formation in the flow field as well as the void fraction at any point in the flow field. Predictions from the model agree very well with our experiments. Model predictions explain differences in the initial location of bubble formation for fuel solutions pre-saturated with CO_2 as opposed to CO_2-free solutions. Experiments with KOH and LiOH added to fuel solutions confirm the validity of the model extension that includes solubility that is enhanced by chemical reaction.

Experiments with LiOH, KOH, and ammonium hydroxide show that the long-term durability of standard Pt-Ru/Nafion/Pt membrane electrode assemblies is compromised because of the presence of lithium, potassium, and ammonium cations that interact with the Nafion membrane and result in increasing the ohmic limitations of the polymer electrolyte membrane. Experiments with $Ca(OH)_2$, while reducing gas formation, precipitate the product $CaCO_3$ out of solution too rapidly for downstream filtering, blocking channels in the flow field.

564b Modeling of Liquid Water in Pem (Proton Exchange Membrane) Fuel Cell

Ulaganathan Nallasivam[1], Debangsu Bhattacharyya[1], Raghunathan Rengaswamy[1], and R. Madhusudana Rao[2]. (1) Chemical Engg. Dept., Clarkson University, 8 Clarkson Avenue, Potsdam, NY 13699, (2) UTC Power, 195 Governor's Highway, South Windsor, CT 06074

A detailed model of a PEMFC is developed considering the effect of liquid water in all the layers of cathode (namely gas flow channel, back up substrate, diffusion layer and catalyst layer). The catalyst layer is characterized through a spherical flooded agglomerate approach in line with experimental evidence. A thin film of ionomer is considered on top of the spherical agglomerate. It is assumed that the water produced in the reaction sites diffuses to the surface of the agglomerate forming a thin film before undergoing evaporation/condensation process. An effectiveness factor approach is adopted to calculate the rate of oxygen reduction reaction. Cathode experimental data for a wide range of flow rates are used to validate the model. The simulation results and the validation with the experimental data depends heavily on the liquid water saturation related boundary conditions. The boundary conditions for liquid water saturation are not well discussed in the existing literature. Depending on the wettabilities of the different materials and the operating conditions, the saturation may vary widely at the boundary. We propose a set of possible boundary conditions in order to capture the physics of the liquid water at the interphase. The model captures the effect of the change in the interphase saturation on the performance of the cell. With the change in the contact angle on the surfaces, the driving force for liquid water flow will also vary significantly affecting the cell power density. Several sensitivity studies will be discussed to assess the impact of the underlying assumptions. The simulations clearly show that the interfacial liquid coverage plays a very important role in determining the limiting current density of the cell in a given operating condition. The use of this model in design and optimization studies will also be outlined.

Oscillatory Flow in the Direct Methanol Fuel Cell: Effects on Mass Transfer, Bubble Management, and Cell Performance

Philip P. Schonewill and David T. Leighton, Jr.
Department of Chemical & Biomolecular Engineering
University of Notre Dame
Notre Dame, IN 46556

The direct methanol fuel cell (DMFC) suffers performance decay at high current densities because of mass transfer limitations at the cell anode. The catalyst instantaneously consumes any methanol which arrives at the surface and the reaction becomes diffusion-limited. Simultaneously, a significant amount of carbon dioxide is generated by the reaction and gas bubbles are formed. The bubbles, if they are not swept out of the fuel cell by the flow, can interfere with the diffusion of methanol to the catalyst by physically blocking diffusion pathways.

In previous work, we have shown that the mass transfer limitations in the direct methanol fuel cell can be overcome by superimposing an oscillatory flow on the steady anode feed. An oscillatory flow which is tangential to the catalyst surface was induced in the porous diffusion layer and enhanced the mass transport of methanol, resulting in improvements of up to 100% for the limiting current density and 30% for the peak power density. The enhancement is observed only if the stroke volume and frequency of the oscillations are chosen such that the Peclet number in the diffusion layer exceeds one. The Peclet number Pe is defined as $Ua(D_o)^{-1}$, where U is the time-averaged fluid velocity, a is the radius of the fibers in the diffusion layer, and D_o is the molecular diffusion coefficient. The conditions of these experiments were such that carbon dioxide gas was not a significant mass transfer inhibitor.

In this work, the operational parameters of a 5 cm^2 direct methanol fuel cell were probed further in order to determine the most advantageous and efficient use of fluid oscillations. The membrane electrode assembly was a commercially available product (anode: 4.0 mg/cm^2 50:50 Pt:Ru, cathode: 2.0 mg/cm^2 Pt, membrane: Nafion 117). The variables examined in this study include the concentration of methanol in the feed, the steady feed rate of methanol/water solution, and the thickness of the diffusion layer. The performance of the cell with and without oscillation was measured to observe its effect.

For example, a thicker diffusion layer, in conjunction with a higher methanol concentration in the feed and superimposed fluid oscillations, was observed to generate a larger peak power density and limiting current density than a cell with a thinner diffusion layer. An example of this is shown in Figure 1. The thicker diffusion layer serves as a barrier which reduces the amount of methanol exposed to the membrane at low current densities, thereby limiting crossover. At higher current densities, the oscillations enhance the mass transfer of methanol through the diffusion layer, which reduces the amount of time the anode reaction is diffusion-controlled.

Significant carbon dioxide bubble generation was also observed at low steady feed rates (*e.g.* 1.0 $cm^3\ min^{-1}$) which rendered the fuel cell current density unstable. The degradation in performance at low flow rates can be partially reversed by symmetric fluid oscillations, but some bubble volume remains to disrupt the diffusion of methanol to the catalyst. In spite of this, performance at lower flow rates – that is, flow rates at which carbon dioxide bubbles are significant enough to interfere with fuel delivery to the catalyst – can approach that of higher flow rates, where carbon dioxide is more readily swept out of the flow field. At even larger flow rates, the bubbles are no longer much of a problem, but the fluid does not reside in the cell long enough to reach the cell temperature. Consequently, the performance can suffer. These three flow rate regimes are illustrated in Figure 2.

If the fluid oscillations are anti-symmetric, the bubbles can be given a net convective displacement out of the cell in spite of the low steady flow rate. Preliminary experiments using anti-symmetric oscillations have been promising. Based on these and other experimentally observed phenomena, recommendations will be made for optimum use of oscillatory flow in the anode of a direct methanol fuel cell, including discussion of the use of novel diffusion layers in the anode of the DMFC.

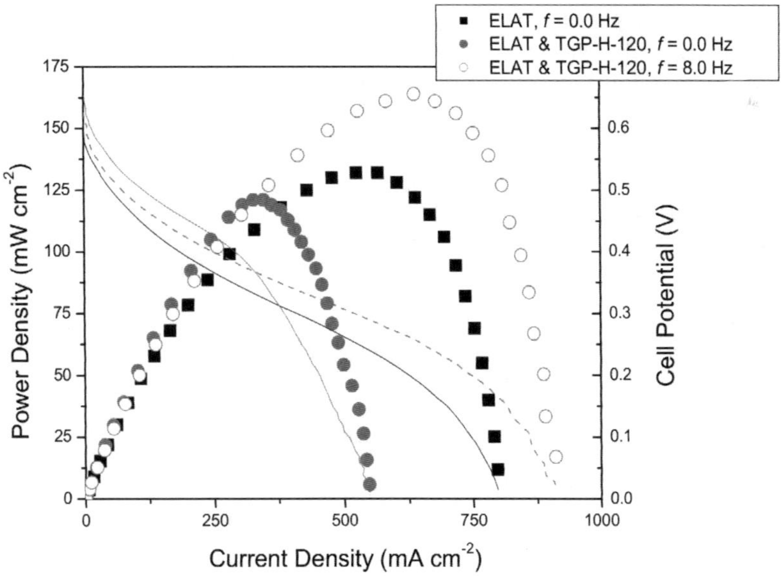

Figure 1. Polarization curves obtained at 70 °C for a thin diffusion layer (ELAT) and a much thicker diffusion layer (ELAT + TGP-H-120) at an anode feed concentration of 2.0 mol L^{-1}. The anode feed rate in each case was 12 $cm^3\ min^{-1}$ and the cathode feed rate was 500 $cm^3\ min^{-1}$. The voltage curves with solid lines correspond to the power curves with solid symbols; likewise, the dashed lines correspond to the open symbols.

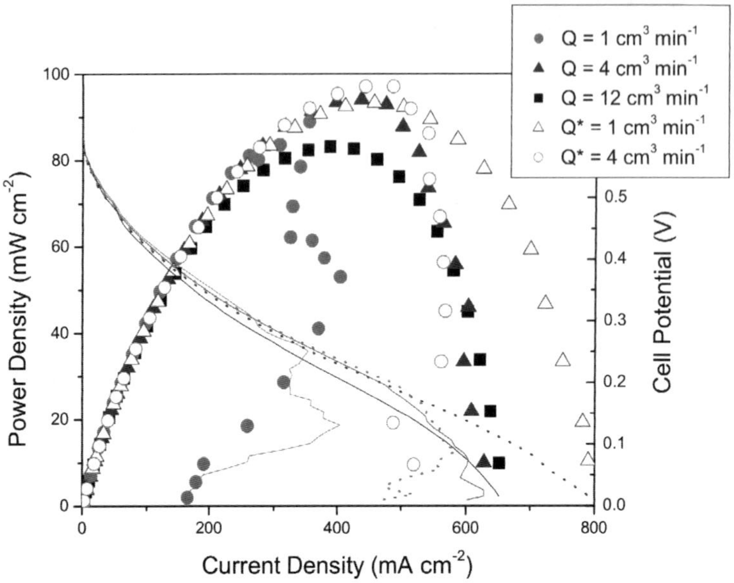

Figure 2. Polarization curves obtained at 70 °C using the ELAT carbon cloth diffusion layer, a cathode feed rate of 500 cm^3 min^{-1}, and an anode feed concentration of 1.0 mol L^{-1}. The voltage curves with solid lines correspond to the power curves with solid symbols and refer to experiments without any fluid oscillations; likewise, the dashed lines correspond to the open symbols and refer to experiments which had an oscillation frequency of 4.0 Hz.

564d A Steady State Three-Dimensional Model for a Pemfc System

Arun Pandy[1], Arvind Raghunathan[1], Subbarao Varigonda[1], Nikunj Gupta[2], and Andrew T. Haug[3]. (1) Systems Department, United Technologies Research Center, 411 Silver Lane, MS 129-15, East Hartford, CT 06108, (2) Shell Hydrogen, Shell Oil Company, 910 Louisiana Street, Houston, TX 77002, (3) UTC Power, 195 Governor's Highway, South Windsor, CT 06074

The automotive industry is making progress toward development and commercialization of PEM fuel cells. PEM fuel cells (PEMFC) are an attractive alternative for the internal combustion engine in automotive applications because they are cleaner and more fuel-efficient. More importantly, automotive fuel cells enable the goals of reduced-dependence-on-oil and transition to a hydrogen economy. In order to successfully design PEMFC systems for these applications development of a thorough understanding of various phenomena that occur during the operation of these systems is needed. Mathematical modeling can be used to improve the understanding of these systems and to help improve the design of these systems.

In this work, a steady state three-dimensional mathematical model that accounts for the coupling between transport phenomena, phase change, and fuel cell electrochemistry in a PEMFC stack is presented. The model captures the effect of multi-component diffusion of gases between the gas channels and the catalysts through the porous layers of the fuel cell using Stefan-Maxwell equations. Pore-flooding as a function of the capillary pressure in each of the porous layers of the unitized electrode assembly is modeled. Water transport through the porous layers is modeled using Darcy's Law. Water transport through the ionomer in the membrane and the catalyst layers is modeled by a combination of pressure driven flow, electro-osmotic drag, and concentration gradient driven back diffusion. The hydrogen oxidation reaction and the oxygen reduction reaction in the catalyst layers is modeled using Butler-Volmer kinetics. In the gas channels the pressure drop associated with two-phase flow is also modeled. This model architecture allows the capability to simulate both single cell and stack systems. Through simulations of this model the effect of different operating conditions – relative humidity, temperature, and pressure of the gases – and system parameters – membrane thickness, gas channel design, etc. – on water management and steady state performance of the fuel cell can be studied. Results from these simulations will also be presented and discussed.

564e Methanol Transport in a Passive Dmfc Employing a Porous Carbon Plate

Mohammad Ali Abdelkareem and Nobuyoshi Nakagawa. Environmental and Chemical Engineering, Gunma University, 1-5-1 Tenjin, Kiryu shi, Gunma ken, 375-8501, Japan

We have demonstrated that a passive DMFC with a porous carbon plate, PCP, significantly reduced the methanol crossover and increase both of the fuel efficiency and the power density of the DMFC. The CO2 gas layer formed between the MEA and the PCP was essential for significantly limiting the methanol transport to the anode surface where the gaseous methanol diluted in the CO2 gas. In this study, the effect of the pore structure and thickness of the PCP, as well as the CO2 gas barrier layer, thickness, pressure, and the discharging route whether through the PCP or not, on the methanol transport and the performance of a passive DMFC were investigated. The main factors for controlling the methanol transport were the barrier of the gas layer with CO2, and the properties of the PCP. At the low current densities of less than 60mA/cm2, when no CO2 bubbles are emitted, both the pore structure and thickness of the PCP did not affect the methanol transport and the current voltage relationship. At the higher current densities, CO2 bubbles were evolved through the PCP, and different resistances to the methanol transport were observed depending on the PCP pore structure and thickness. The discharging of CO2 bubbles across the PCP enhanced methanol transport in case of large pore diameter PCP, 42 micro meter in contrast to the small pore diameter one, 1micro meter. The resistivity of the CO2 gas layer increased by increasing its thickness while it not affected by the variation of its pressure in case of PCP with small pore diameter within the measured pressure range.

610a Validation of Les Based Predictions of Heat Flux to Objects in Transportation Fuel Fires

Jennifer Spinti, Philip J. Smith, Jeremy N. Thornock, and Stanislav Borodai. Chemical Engineering, University of Utah, 155 S. 1452 E. Rm 380, Salt Lake City, UT 84112

Combustion simulation has made great strides recently with the successful use of Large Eddy Simulation (LES) to simulate systems where unsteady effects are important. An example where such effects are important is in large pools fires. For example, the puffing nature of the fire is the result of periodic, large-scale air entrainment caused by the formation of small scale Rayleigh-Taylor instabilities at the base of the pool (Tieszen et al. 2004). The nature of the air-entrainment behavior affects the properties of the flow field, including the locations of the reaction zones and the radiative heat transfer to objects in or near the fire. Thus, for accurate prediction of the flow field along with the accompanying features of the flow field, one must capture the unsteady effects in the simulation. In this discussion, we will focus on the use of an LES tool, called ARCHES, to compute heat flux to objects in or near large fires. We employ the Validation and Verification methodology of Oberkampf and coworkers to assess the quality of the various submodels within ARCHES by using experimental data germane to each submodel. We also use experimental data from Greiner and coworkers (University of Reno) and Blanchat and coworkers (Sandia National Labs) to validate hear flux to objects in large transportation pool fires. Utilizing validation metrics, we assess the level of agreement between the ARCHES LES simulation and the experimental measurements.

HEAT AND MASS TRANSPORT IN A REACTIVE POROUS MEDIUM: APPLICATION TO OIL SHALES AND HEAVY OIL COMBUSTION

A. Lapene
Total - Exploration & Production Pau - France / Institut de Mécanique des Fluides de Toulouse - UMR 5502 - Toulouse - France

B. Debenest, M. Quintard
Institut de Mécanique des Fluides de Toulouse - UMR 5502 - Toulouse - France

M. F. Martins, S. Salvador
Ecole des Mines d'Albi-Carmaux - Centre Rapsodee - UMR CNRS 2392 - Albi - France

Introduction

During the past few years, prices of crude petroleum have increased very sharply, with the price rising from 11 dollars per barrel in february 1999 to 75 dollars per barrel last august. This rising starts to make profitable the exploitation of heavy oil difficult to extract compared to crude oil or coal. Oil shales and heavy oils have been studied during the Eighties but given up during Nineties for more profitable sources of energies. Now, regarding barrel price they become again interesting. Oil shales are sedimentary rocks with a significant part of organic matter called kerogen. The kerogen can be converted to oil through pyrolysis between 400 and 500℃. Oil shale can also be burned directly as a low-grade fuel. The United States Energy Information Administration estimates the world supply of oil shales at 2.6 trillion barrels of recoverable oil. One of a converting method is to burn a part of them to pyrolyze the other with the generated heat. This method can be applied in-situ or in a reactor after mining extraction. Compared to other materials, the study of oil shales is fairly difficult. From the chemical point of view, they are more complex than coal, for example, because they contain various organic components. However, the presence of a large fraction of inert inorganic matter in the rock induces a big simplification since the combustion does not affect significantly the bed structure. Crude oil is considered "heavy" if its API gravity is lower than 20 degrees and its viscosity is higher than 1000 centipoises. Heavy oils are difficult to exploit because of their low gravities which correspond to high viscosities. Thus, it can prevent their ability to flow within a reservoir. Moreover they are often mixed with sand or other minerals. We can find some different methods to exploit them. One of them is to directly inject solvent in the well to recover oil. Another way is to inject hot water to lower its viscosity. Also, like for oil shales, we can use the In-situ combustion method.

We are interested by modeling in-situ combustion. To get some relevant informations, combustion tubes are usually used at the laboratory scale. Families of oxidation reactions are the same ones for both materials but the study of oil shales is easier because few water is present inside the material; thus , we don't take into account the water phase. Moreover, for the heavy oil, coke is due to the fuel deposition through oil vaporization and cracking of the heavy components of the oil (Mamora (1993)) while for oil shales coke is only due to cracking. In this paper, we will deal with modelling oil shales combustion. After this introduction we will finish with a discussion on additional difficulties of modelling heavy oil combustion and similarities with oil shales.

Modelling of oil shales combustion

Numerical modelling of heterogeneous combustion of fuel in fixed beds must correctly describe coupled mass and heat transport in a reactive porous medium. These phenomena are strongly coupled and non-linear. There are two visions to deal with this problem: a macroscopic vision and a microscopic vision.

Microscopic vision

In the literature, there are only few authors who attempted numerical simulation at a microscopic scale. Ohlemiller *and al.* (1980) gave a very detailed review of the chemical and physical processes implied in fixed-bed solid fuel combustion. They proposed a complete mathematical model taking into account all transport mechanisms and all complex chemical processes at grain scale. Then they estimated that the complete problem was not approachable. Lu and Yortsos (2000) presented numerical simulations where the porous medium had been represented by capillaries network. Work is done at the pore scale in off-line solids with a local approach for transport. Various transfers and couplings were modelled using effective coefficients, consequently, this cannot be considered like a true microscopic approach. Redl (2002) come near this objective. He used a Lattice-Boltzmann type of approach in a three-dimensional space. He solved transport equations in each phase at the microscopic scale. However, oxidation is treated like a homogeneous reaction occurring into the liquid phase, which cancels the majority of local couplings between transport and reactions. Debenest *and al.* (2005) adopted an original approach which was based on a detailed three-dimensional description at the microscopic scale. Thus, they could carry out direct simulations in order to characterize various modes of propagation and to determine transport effective parameters at the Darcy-scale.

Macroscopic vision

In most cases, a macroscopic point of view is adopted, whereby the medium is regarded as a continuum. The equations involve various effective parameters, such as dispersion coefficients or mean volumetric reaction rates, and they relate macroscopic fields that represent volume averaged local quantities. However, the effective coefficients depend both on the medium microstructure and on the flow and reaction regimes. Their actual determination generally requires the solution of the full featured set of microscale equations in a representative sample (Adler *and al.* (1992) and Adler *and al.* (1998)). In addition, the validity of homogenized equations is also questionable, in view of the very strong gradients at the microscopic scale. For instance, in the reference lab experiment, the reaction front is only a few grain sizes thick, and the temperature jumps by several hundreds of degrees over this distance. Finally, local non-equilibrium plays an important role in many respects, such as in pollutant formation, and it can only be modelled by detailed local simulations.

Strategy

We will model coupled mass and heat transport in reactive porous medium using an homogeneous description at the Darcy-scale. Local non equilibrium transport of heat will be treated with a two field temperature model, one for the gas phase and one for the solid phase. The used reaction mechanisms will come from various experiments. Firstly, we will consider the most simple oxidation mechanism with only one reaction as we can see in the "oil" approach. Finally, a two reaction mechanism will be used to take into account initial conditions variations like oxygen partial pressure at the inlet.

Experiments

Fixed Bed Experiment

This experiment has been carried out in a fixed bed reactor with co-current air flow. The porous medium is made of a granular medium constituted by crushed oil shale (75%) and sand (25%). Grain diameter is between 500 and 1000µm. The reactor is a cylinder with an external diameter of 90mm and internal diameter of 54mm (Figure 1). Twelve thermocouples organized in two groups permit to measure temperature in different zones. The first group (T1, T2, T3, T10, T11 and T12) is laid out along the reactor, on the cylinder symmetry axis, each one being separated by 45mm. The second group (T4 to T9), takes place on a horizontal perimeter, near the walls, and permits to estimate the radial thermal losses. During experiments, we can sample gas in the reactor core thanks to a new micro-sampling device developed by Martins and al. We will essentially be interested by carbon monoxide and carbon dioxide measurements.

Thermo-gravimetric analysis (TGA) and Differential Scanning Calorimetry (DSC)

Two experiments have been carried out, each one with TGA and DSC measurements: the first one with nitrogen atmosphere (Figure 2a) and the last one with oxidizing atmosphere (Figure 2b). In nitrogen atmosphere experiments, we can see two typical oil shale pyrolysis mechanisms. The first one acts on the oil shale organic matter while the second one acts on carbonates and is therefore named decarbonatation. This last one is very endothermic whereas the first one can be considered approximately as athermic. We suppose that each reaction is first order and follows an Arrhenius law. Thanks to a graphical method, we can obtain the frequency factor and activation energy for each reaction. The results are shown in Tab. (1). In oxidizing atmosphere, in addition to pyrolysis we can see two oxidation families extremely exothermic in the form of two peaks in the DSC results. The first peak corresponds to "Low Temperature Oxidations" (LTO) performed on oil shale light components. the second peak corresponds to "high temperature oxidation" (HTO) performed on the residual solid called "coke". This vocabulary is borrowed from the chemical petroleum literature.

Modeling of Mass and Energy Transport

Many approximations have been made in this first study. However, we will quote only some of them. Initially, we consider a one dimensional problem. This assumption is based on the reactor geometry, in particular its symmetries, but the main argument is that the observed flow is mostly one dimensional. The other major assumption relies also on experimental observation: during and after combustion, the bed structure is little affected. So, porosity can be taken as a constant. Considering all the approximations, we may write the various balance equations as described below. The first is the mass conservation equation

$$\varepsilon \frac{\partial \rho_g}{\partial t} + \frac{\partial}{\partial x}(\rho_g v) = R_g \qquad (3)$$

In this equation, ε, ρ_g, t, x, v, R_g are respectively the gas porosity, gas density, time, position, filtration velocity and total mass source term for species in gas phase. The second equation is the species evolution equation for species *k* in the solid phase

$$\varepsilon\rho_s \frac{\partial Y_{k,s}}{\partial t} = R_{s,k} \tag{4}$$

In this equation ρ_s, $Y_{k,s}$ and $R_{s,k}$ are respectively the solid density, mass fraction of species *k* in the solid-phase and mass source term of species *k* in the solid-phase. The third equation is the species evolution equation for species k in the gas-phase

$$\varepsilon_g \frac{\partial \rho_g Y_{k,g}}{\partial t} + \frac{\partial}{\partial x}\left(\rho_g v Y_{k,g}\right) - \frac{\partial}{\partial x}\left(\rho_g D_k \frac{\partial Y_{k,g}}{\partial x}\right) = R_{g,k} \tag{5}$$

In this equation $Y_{k,g}$, D_k and $R_{g,k}$ are respectively the mass fraction of species *k* in the gas-phase, effective diffusion coefficient of species *k* and the mass source term of species *k* in the gas-phase. The forth equation is Darcy's equation

$$v = -\frac{K}{\mu}\frac{\partial P}{\partial x} \tag{6}$$

where we have neglected gravity effects. In this equation K, μ and P are respectively the permeability, dynamic viscosity and pressure. The fifth and sixth equations are the heat equations

$$\varepsilon_g (\rho C_p)_g \frac{\partial T_g}{\partial t} + (\rho C_p)_g v \frac{\partial T_g}{\partial x} = \frac{\partial}{\partial x}\left(\lambda_g \frac{\partial T_g}{\partial x}\right) + \Gamma(T_g - T_s) + Q_g + h(T_{amb} - T_g) \tag{7}$$

$$\varepsilon_s (\rho C_p)_s \frac{\partial T_s}{\partial t} = \frac{\partial}{\partial x}\left(\lambda_s \frac{\partial T_s}{\partial x}\right) + \Gamma(T_s - T_g) + Q_s + h(T_{amb} - T_s) \tag{8}$$

There is one equation for each phase in order to take into account local non-equilibrium aspects (see, for instance, Quintard *and al.* (1997) and Quintard *and* Whitaker (2000)). In these equations, C_p, T_g, T_s, λ_g, λ_s, Γ, Q_g, Q_s, h and T_{amb} are respectively the heat capacity, temperature of gas phase, temperature of solid phase, effective thermal conductivity of gas phase, effective thermal conductivity of solid phase, exchange term between both phases, heat source term in gas phase, heat source term in solid phase, thermal exchange coefficient and ambient temperature. Thermal losses are taken into account through the *h* coefficient. The seventh and last equation is the perfect gas equation

$$\rho_g = \frac{P M_g}{T_g R} \tag{9}$$

In this equation M_g and R are respectively mean molar weight and perfect gas constant. Finally, we may combine Eq. (3), Eq. (6) and Eq. (9) to obtain the following pressure equation

$$\frac{\varepsilon M_g}{RT_g}\frac{\partial P}{\partial t}-\frac{\varepsilon M_g P}{RT_g^2}\frac{\partial T_g}{\partial t}+\frac{\varepsilon_g P}{RT_g}\frac{\partial M_g}{\partial t}+\frac{\partial}{\partial x}\left(-\rho_g\frac{K}{\mu}\frac{\partial P}{\partial x}\right)=R_g \qquad (10)$$

Chemistry Modeling

The decarbonatation reaction is modeled by the following equation

$$CaCO_3 \rightarrow CO_2 + CaO \qquad (11)$$

Carbonates initial quantity available is determined by TGA and enthalpy of reaction by DSC (we found 441 J/g). The major modeling problem was to obtain a robust and correct kinetic scheme which matched the experimental results. We have chosen three criteria to construct our reaction mechanism and it parameters: front velocity, maximum front temperature and $[CO_2]/[CO]$ ratio at outlet. Our first attempt was to use the coal model from Zajdlik *and al.* (2001) considering that fuel was char after pyrolysis. This model comprises four equations, three are heterogeneous and one is homogeneous. However, this model does not correctly represent experimental data. Indeed, an activation energy that is too high does not permit to ignite combustion like experiments. Moreover, reaction enthalpies are not compatible with the front velocity and maximum front temperature. In petroleum literature and particularly in in-situ combustion literature, we can find simple reaction mechanism which correctly matches experimental results. We use the following simple scheme

$$Fuel+\frac{2\alpha+1}{2(\alpha+1)}O_2 \rightarrow \frac{1}{1+\alpha}CO+\frac{\alpha}{1+\alpha}CO_2 \qquad (12)$$

In this equation α is the $[CO_2]/[CO]$ oxidation reaction ratio. We have to notice that α is not equal with the $[CO_2]/[CO]$ ratio at outlet because decarbonatation reaction produces CO_2. The initial fuel quantity available is calculated assuming Eq. (13) stoichiometry and calculating consumed oxygen. Another way is to mix the "coal" and "petroleum" views. We keep Eq. (13), we impose the value of α to be zero and we add a reaction which oxides CO to CO_2

$$CO+\tfrac{1}{2}O_2 \rightarrow CO_2 \qquad (13)$$

Then, the outlet $[CO_2]/[CO]$ ratio is given by Eq. (13) and it value depends on the oxidant presence and, by consequence, depends on the inlet oxygen partial pressure.

Results

Confrontation with experimental results

We have simulated the fixed bed experiments as close as possible to the experimental protocol. Figure (3a) shows simulated results for both reaction models together with the experimental results. Front velocity, maximum temperature, temperature increase and temperature decrease fairly match. The one reaction model allows a faster temperature rise than the two reactions model. Indeed, the homogeneous reaction activation energy is higher than the one of the heterogeneous reaction and becomes significant later when temperature is higher. In both models, we can see that temperature levels are lower than in experiments. Fuel oxidation reaction (LTO and HTO) are lumped into only one reaction. However LTO reactions are known to play a major role in combustion ignition. In our model we can only account for a "mean reaction".

Parameters sensibility and major differences between both models

In terms of temperature profiles, both models are similar. We are now interested by visible differences between both models. In this goal we have studied three configurations by changing the initial injection velocity. Three injection velocities were chosen. The first is the same that in the experiment (3.6×10^{-2}m/s) and correspond to a mass transport Peclet number (based on granular diameter) of 2, the second is half the experiment velocity and the third is double of the experiment velocity. The resulting mass transport Peclet numbers are equal to 1, 2 and 4. We will look at the maximum front temperature for the gas in Fig. (3b), the front velocity in Fig. (3c), and the $[CO_2]/[CO]$ ratio in Fig. (4d). The abscissa coordinate, t^*, is a dimensionless time defined as $t^* = t/t_{out}$ with t_{out} the time at which the combustion front leaves the reactor. Maximum gas temperature and front velocity evolutions are almost the same for both models. However temperature rise is slightly faster for the one reaction model. The main discrepancy between both models is about gas production. For the experimental injection velocity, both models permit to obtain approximately the same value but in the two reactions model this ratio is more depending upon time, especially at the beginning. For other injection velocities, both models give very different results. The ratio for the one reaction model is always the same with little discrepancies, whereas, for the other model, it depends highly on the injection velocity. Indeed, in the one reaction model, the $[CO_2]/[CO]$ ratio is fixed, for the heterogeneous reaction, and little variation are due to decarbonatation reaction whereas, in the other model, it is controlled by a homogeneous reaction and by consequence by temperature and oxygen partial pressure through Arrhenius law.

Heavy oil study

Modeling and numerical aspects

There are important discrepancies between oil shales combustion and heavy oil combustion. Indeed, study of heavy oil combustion is more difficult. First difference is that three mobile and miscible phases must be taken into account. By consequence, mathematical model become more complex with the addition of one saturation equation by phase. To envisage a good fuel deposition we have to model mass exchange between phases with a good accuracy. In other words, we have to choice a good equation of states for each phase (Peng *and* Robinson (1976)) and calculate right equilibrium states thanks to thermodynamics laws through equilibrium constant. Some famous papers present complete mathematical models (Aziz *and* Settari (1979), Coats (1980)). Phases equilibrium is usually compute with a flash calculation (Michelsen (1982)). In addition to transport and thermodynamic aspect we

have to treat chemical reactions. We know that there are similar to those which occur for oil shales combustion. Numerically, an ODE's system has to be solved.
Considering all previous points, a complex non-linear mathematical system is obtained and it is very tricky to compute it. Usually, an operator splitting technique is used.

Reactional mechanism

Before modeling at the combustion tube scale, we have to experiment at the kinetic tube scale. The kinetic tube consisted of a thick-walled stainless steel cylinder measuring approximately 10 cm long with an O.D. of 5 cm. It is filled by sand and oil and temperature is linearly increased whereas air is injected. It permits to determine and validate a reactional mechanism and to evaluate its kinetic parameters. Thanks to oil shales experiments a mechanism has been intuit but past experiments with heavy oil has shown that the water influence chemistry. Indeed, we have to design original experiments to evaluate this influence. They are in hand.

Conclusion

We have developed a numerical tool reproducing correctly experiments results for oil shale combustion. The main difficulty was to find a good reaction mechanism. We have proposed two types: one with only one heterogeneous reaction, the simplest view, and another with a heterogeneous reaction and a homogeneous reaction. Both models gave same results compared with experimental thermal probes and gas analysis if we fixed the experimental conditions. If we change, for instance, the injection velocity, gas production is highly dissimilar. We do not know which view is the right one but we have proved that, if the temperature profiles are the same, the gas production is clearly modified. It seems important not to fix α value because if we change, for instance, the initial fuel quantity, modifying the sand quantity, the results will be also modified. Moreover, we indicated that LTO reaction is not considered yet. We must add this reaction to allow a better ignition at low temperatures. Finally, thermal losses are taken into account only with a simple exchange term and a two dimensional program would be an interesting tool to investigate the influence of this point. All these observations permit us to focus on some important aspects for the heavy oil combustion. We have to carry out an accuracy analysis of outlet gases and particularly CO and CO_2. Our new numerical program will have to be 2D tool and we are also going to introduce a three phase transport model. Reactive and thermodynamics mechanisms will have to be also taken into account with the influence of water content in the gas phase.

Tables

Table 1. Activation energy and frequency factor for both pyrolysis

	Frequency factor	Activation energy
Pyrolyse I	39.30 s^{-1}	75490 J
Pyrolyse II	3.9×10^{6} s^{-1}	199369 J

Graphics

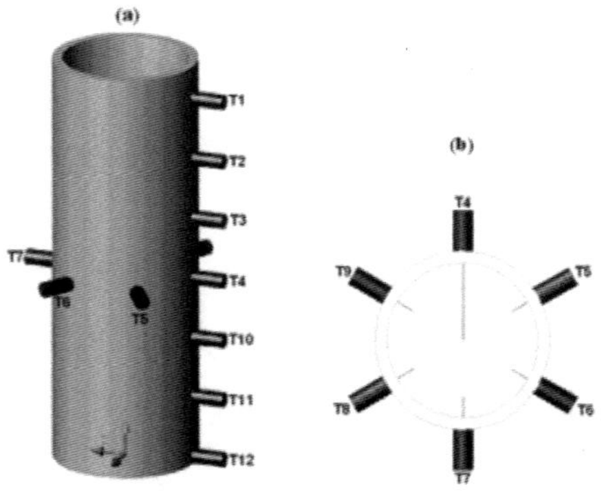

Figure 1. Reactor 3D view (a) and top view (b)

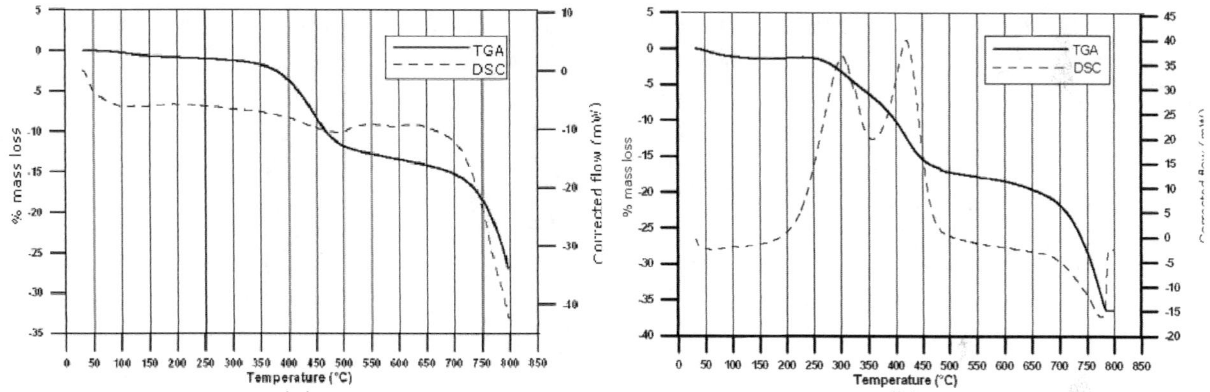

(a) (b)

Figure 2. TGA and DSC with nitrogen atmosphere (a) and oxidizing atmosphere (b) with heating-rate of 3℃/min

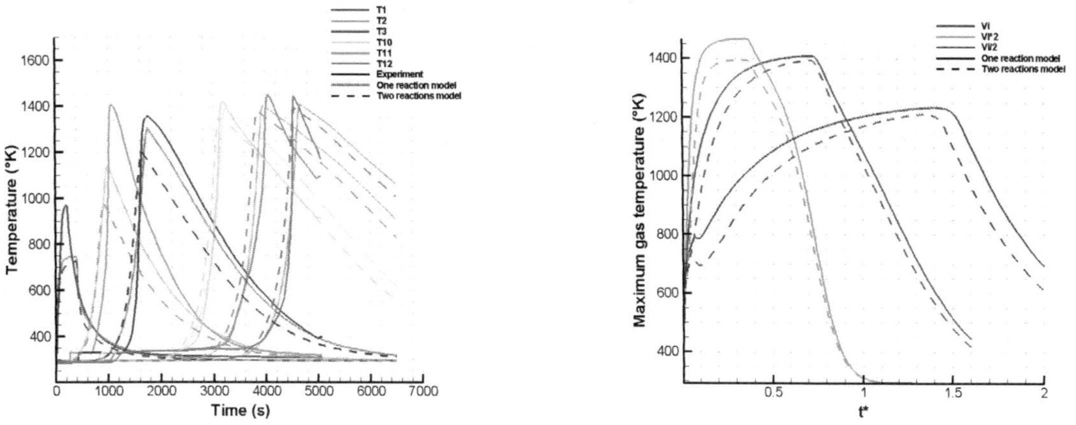

(a) (b)

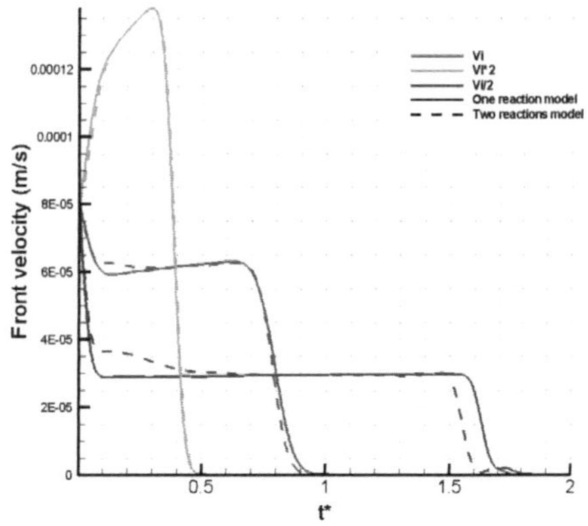

(c)

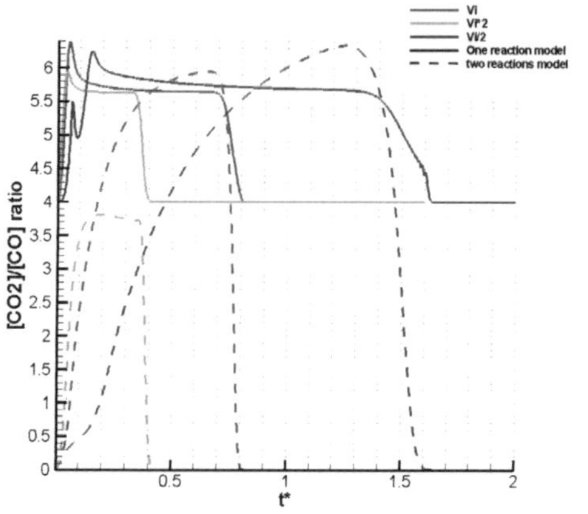

(d)

Figure 3. Comparison between experiment and simulation (a) and between both reactional mechanisms (b), (c) and (d)

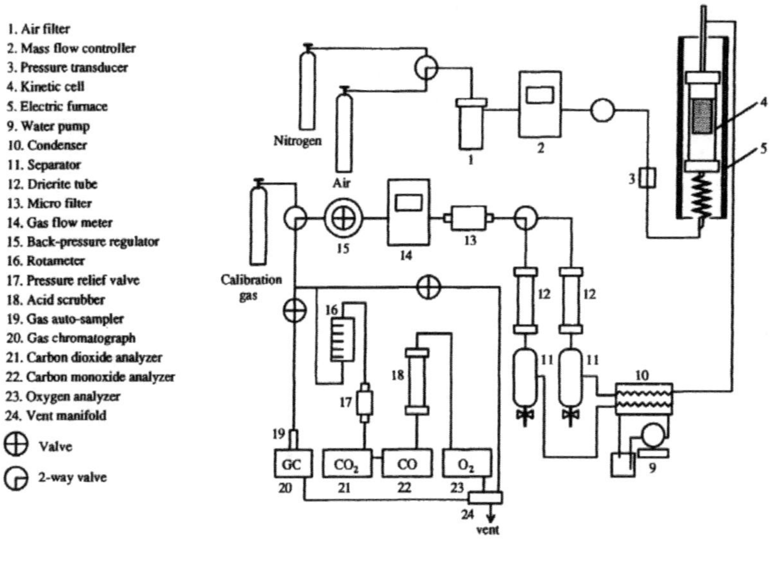

Figure 4. Schematic diagram of kinetic tube and apparatus (Mamora (1993))

Acknowledgements

We wish to thank Total s.a for his financial support on the heavy oil part and J.F. Thovert, V. Mourzenko and M. Elayeb for their contribution.

References

1. D. D. Mamora, PhD Thesis, Kinetics of in-situ combustion, (1993).
2. P.M. Adler, Porous Media: Geometry and transports, Butterworth-Heinemann (1992).

3. P.M. Adler and J.F. Thovert, Real porous media: local geometry and macroscopic properties, Appl. Mech. Rev., Vol. 51, 537-585 (1998).
4. G. Debenest, V.V. Mourzenko and J.F. Thovert, Smouldering in fixed beds of oil shale grains. A three-dimensional microscale numerical model, Combustion Theory and Modelling, Vol. 9, 113-135 (2005).
5. G. Debenest, V.V. Mourzenko and J.F. Thovert, Smouldering in fixed beds of oil shale grains: governing parameters and global regimes, Combustion Theory and Modelling, Vol. 2, 301-321 (2005).
6. C. Lu and Y. Yortos, The dynamics of combustion in porous media at the pore-network scale, ECMOR VII (Baveno, Italie, Septembre 2000).
7. T.J. Ohlemiller, Modeling of smoldering combustion propagation, Prog. Ener. Combust. Sci., Vol. 11, 277-310 (1985).
8. C. Redl, In situ combustion modeling in porous media using lattice boltzmann methods, ECMOR VIII (Freiberg, Allemagne, Septembre 2002).
9. M. Quintard, M. Kaviany and S. Whitaker, Two-Medium Treatment of Heat Transfer in Porous Media: Numerical Results for Effective Properties, Advances in Water Resources, Vol. 20, 77-94 (1997).
10. M. Quintard and S. Whitaker, 2000, Theoretical analysis of transport in porous media, Handbook of heat transfer in porous media. Marcel Decker Inc., N.Y., 1-52 (2000).
11. R. Zajdlik, L. Jelemensky, B. Remiarova, and J. Markos. Experimental and modelling investigations of single coal particle combustion. Chemical Engineering Science, Vol. 56, 1355–1361 (2001).
12. D.Y. Peng and D.B. Robinson, Ind. Eng. Chem., Fundam., Vol. 15, No. 1, 59-64 (1976)
13. K. Aziz and A. Settari. Petroleum Reservoir Simulation. Applied Science Publisher, (1979).
14. K.H. Coats. In-situ combustion model. SPE, pages 533–554, (1980).
15. M. Michelsen. The isothermal flash problem : Part II : Phase split calculation. Fluid Phase Equilibria, 9 :21–40, (1982).

610c Application of Adsorption in Refrigeration by Sulfurized Activated Carbon

Yolanda Pliego-Bravo, Jacqueline Montaño-Äguila, and Eusebio Bolaños -Reynoso. DEPI, Instituto Tecnológico de Orizaba, Av. Oriente 9 No. 852, Orizaba,Veracruz, Mexico

The populational growth and the energy crisis have caused serious problems to the world environment and the sustainable development. The refrigeration for compression of vapor operated with electric power faces the challenge that the chlorofluorocarbons and the hydro chlorofluorocarbons do not favor the sustainable development. Natural refrigerants as water, ammonia, methanol, etc., they will be wellcome in the future of the industry of the refrigeration and the conditioned air. However the good use of this energy should make sure. In this investigation the acting of sulfurized activated carbon was evaluated in the refrigeration cycle for adsorption. The carried out methodology was the following. Starting from the bamboo species Bambusa vulgaris striata, the sulfurized activated carbon was elaborated and characterized. The operation of a refrigeration prototype for adsorption was conditioned and stabilized, a design of experiments bifactorial was applied to evaluate the effect of three activated carbons: a commercial activated carbon, a sulfurized activated carbon prepared starting from the commercial activated carbon, a sulfurized activated carbon starting from bamboo and the effect of two adsorbats: ethanol and methanol. The studied answer variables were temperature in the evaporator, regeneration temperature, adsorbed percentage, percentage desorbed and yield coefficient (Cop). The obtained results were: in the elaboration of sulfurized activated carbon starting from bamboo the following yields obtained in the carbonization 25.6%, in the activation 73.55% and in the sulfurization 79.5%. The refrigeration prototype was conditioned and stabilized the recurrent operation. The results of the statistical analysis of the data for the temperature in the evaporator, the significant factors were the adsorbent and the adsorbat, for the percentage of adsorption the only significant factor is the adsorbat, for the desorption percentage the only significant factor was the adsorbent, for the regeneration temperature the only significant factor was the adsorbat, for the yield coefficient the significant factors were the adsorbent and the interaction adsorbent-adsorbat. For the characterization of the microporosity was applied the equation of Dubinin-Raushkevich and the equation of Dubinin-Astakhov obtaining the best adjustment in 0.9569 for the couple of sulfurized activated carbon, lowest adjustment presented the even sulfurized activated carbon and methanol with 0.8. For the characterization of the microporosity using the equation of Dubinin-Astakhov the best yield was obtained for the system of sulfurized activated carbon and methanol with 0.9559 and the lowest adjustment was for the couple of commercial activated carbon ethanol 0.829. With the obtained results it is concluded: the sulfurized activated carbon increases the value of the Cop= 0.34 of the refrigeration prototype as the best value when it uses methanol as adsorbat.

Numeric Simulation of Radiation Chamber in Furnaces

#610 - Transport Processes in Energy Systems II (07A03)

Eduardo Vila Real Mendes, Prof. D.Sc. Nide Fico Jr., D.Sc. Carlos Eduardo Fontes Silva, D.Sc. Valter Cesar de Souza, Patrícia Carneiro dos Santos

1) Abstract

The computational analysis proposed has the objective of validating physical model of radiation chamber inside furnaces. The main phenomena involved are: energy transport, turbulence, mass transport and chemistry reaction. This study can be used by other researchers as a guideline to choose a model that represents adequately the physical phenomena of great industrial interest. In this paper one has used a commercial software called PHOENICS, from CHAM (www.cham.co.uk), to work out the necessary CFD simulations.

The radiation model used in this study was IMMERSOL, which is a proprietary radiation model developed by CHAM. Combustion phenomena were modeled using the 3 gases model, which considers the reaction fuel + oxidant -> product (only 3 species).

Finally, the temperature, velocity and pressure profiles, among others, will be shown throughout the radiation chamber. Hence, this work aims the study of the main physical phenomena that takes place inside the radiation chamber of a general furnace using computational fluid dynamics (CFD) as a tool.

2) Introduction

Computational Fluid Dynamics (CFD) has been mainly used in phenomena transport research. It has been largely used in energy, automotive and aeronautic industry. With this tool, an engineer can design a machine with more accuracy or solve optimization problems in less time. Another advantage of using CFD is the achievement of high performance while studying a specific transport phenomenon.

Numeric simulations have many advantages in regard to experimental analysis. Some of them are lower costs, possibility of studying scenarios with specific boundary conditions and results availability throughout the solution domain, besides providing faster results.

At present, furnace designs are most based on previous successful projects. Although there are usually few errors by this method, many opportunities of improving efficiency remain unknown. Furnace efficiency is an important variable for industrial process efficiency. Thus improvements in furnace mechanism can reduce production costs and therefore increase industrial performance. So, CFD is an important tool for evaluating industrial processes performance and should be used for development of new furnaces, allowing changes in their design.

3) Mathematical Models

Main governing equations in solving problems by CFD method are briefly explained in next topics.

3.1 Conservation equation

Physical principle: mass can be neither created nor destroyed. Equation (1) shows it mathematically [1]:

$$\frac{\partial \rho}{\partial t} + \nabla \cdot \left(\rho \vec{V}\right) = 0 \tag{1}$$

Where:

ρ : density;

t : time;

V : velocity.

3.2 Momentum equations

Physical principle: force is equal to the changes of momentum in a time rate. Equation (2) shows x-component of momentum equation. [1]

$$\frac{\partial(\rho u)}{\partial t} + \nabla \cdot \left(\rho u \cdot \vec{V}\right) = -\frac{\partial p}{\partial x} + \rho f_x \tag{2}$$

Where:

ρ : density;

V : velocity;

u : velocity in the x-direction;

p : pressure;

f_X : x-component of net body force per unit mass.

Momentum equations for y and z directions can also be deductible.

3.3 Energy equation

Physical principle: energy can be neither created nor destroyed; it can only change in form. In equation (3) this principle is sketched mathematically. [1]

$$\frac{\partial}{\partial t}\left[\rho\left(e + \frac{V^2}{2}\right)\right] + \nabla.\left[\rho\left(e + \frac{V^2}{2}\right)\vec{V}\right] = \rho.\dot{q} - \nabla.\left(p\vec{V}\right) + \rho.\left(\vec{f}.\vec{V}\right) \tag{3}$$

Where:

ρ : density;

e : internal energy per unit mass;

$\dot{q}$: volumetric rate of heat addition per unit mass;

V : velocity;

p : pressure;

f : net body force per unit mass.

3.4 Radiation model: IMMERSOL

IMMERSOL is a radiation model present in PHOENICS software. In this method, a temperature variable T3 is created due to thermal radiation. In a solid region, T3 equation is given by equation (4). [2]

$$c_p \cdot \frac{\partial T_3}{\partial t} - \nabla \cdot (\lambda \cdot \nabla T_3) = \dot{Q} \tag{4}$$

Where:

Cp : heat capacity of the solid;

λ : thermal conductivity of the solid;

$\dot{Q}$: volumetric rate of heat addition.

Calculation of T3 between solid surfaces is governed by equation (5). [2]

$$\nabla \cdot (\gamma_{rad} \cdot \nabla E_3) = (a+s) \cdot (E_{12} - E_3) \tag{5}$$

Where:

E_3: radiosity in solid surface;

E_{12}: mean radiosity between surfaces 1 and 2;

a : absorption coefficient;

s : scattering coefficient;

$$\gamma_{rad} = \frac{4}{3}\left(a + s + \frac{1}{WGAP}\right)^{-1}$$

$WGAP$: distance between surfaces 1 and 2.

3.5 Combustion model: 3-gases

This combustion model is also part of PHOENICS. It is based on the following reaction:

FUEL + OXIDANT → PRODUCT

Thus, reaction occurs without intermediate products and is irreversible. Combustion enthalpy for this reaction is computed by equation (6).

$$h = c_p \cdot T + H \cdot m_{fu} \quad (6)$$

Where:

c_p : heat capacity with constant pressure;

T : inlet temperature reaction;

H : heat of combustion;

m_{fu} : fuel mass ratio not reacted.

4) Problem description

Furnace application in process industry has been important since its beginnings, in the 18th century. Due to industrial revolution, furnaces were largely applied in metallurgic industry and later in oil industry. In figure (1) are shown typical uses of furnaces in process industry. In metallurgy, furnace is of great significance in production of iron, steel and aluminum alloys – figure 1(a) [3]. In oil industry, furnaces are used in several processes that require feed stream heating such as atmospheric distillation - figure 1(b) [4].

(a) (b)

Figure 1 – Furnaces and its industrial application

Main furnace components studied in this research are chimney, convection section, coil, radiation section and burners. These components are sketched in figure (2) and concisely described later. [5]

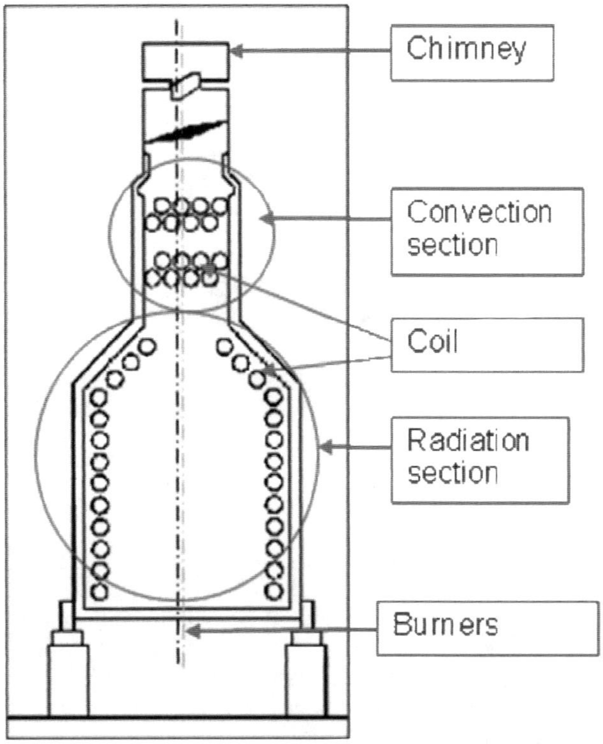

Figure 2 – Main components of a typical blast furnace

Radiation section: in this region occurs fuel combustion through burners. The process coil is constructed at this region and main heat transfer in this section is by radiation.

Convection section: a rectangular box where convection coil is placed.

Chimney: leads the combustion products to atmosphere.

Burners: where combustion takes place.

Coil: internal duct in furnace that allows feed stream heating.

First step for solving a problem with a CFD method is implementation of both geometry and its computational grid in the software. Choosing a grid requires experience and involves machine performance, results accuracy and geometry complexity. Geometry considered in this paper is shown in figure 3. This study used Chen-Kim as its turbulence model, a variation of K-ξ model and set fixed temperature for the tubes.

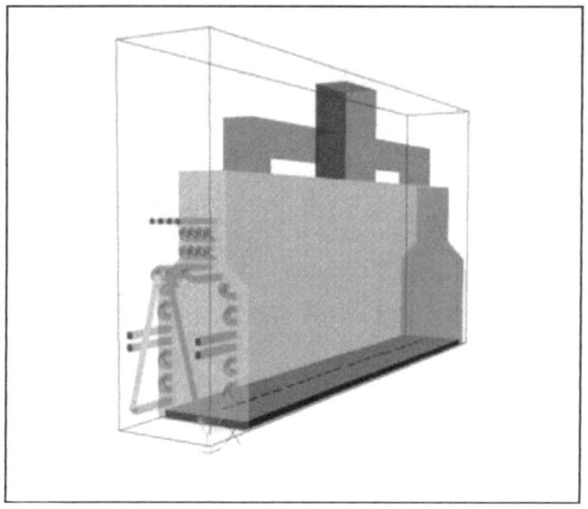

Figure 3 – Geometry used in this CFD analysis

5) Solution Method

The CFD solution method adopted in the present work consisted on simulating a base case in many scenarios. First, combustion and radiation models were not considered. This step is important to evaluate initial results obtained from CFD software because it allows changes in geometry, mesh and boundary conditions with less computational effort since radiation and combustion methods are not included. After defining simulation scenario and obtaining converged results from simulation, researcher knows which are expected results of the system and can analyze contribution of including combustion and radiation models to the simulation. Next step is to include radiation and combustion models and then to be certificated that the simulation is converged. Once base case had already obtained converged results, it will take less time to have good results from this new simulation scenario. If these models were implemented in the beginning before achieving convergence, surely time necessary for having good results would be longer.

6) Results Analysis

As discussed previously, first it was defined a base case without radiation and combustion models. Figure below shows results from this study.

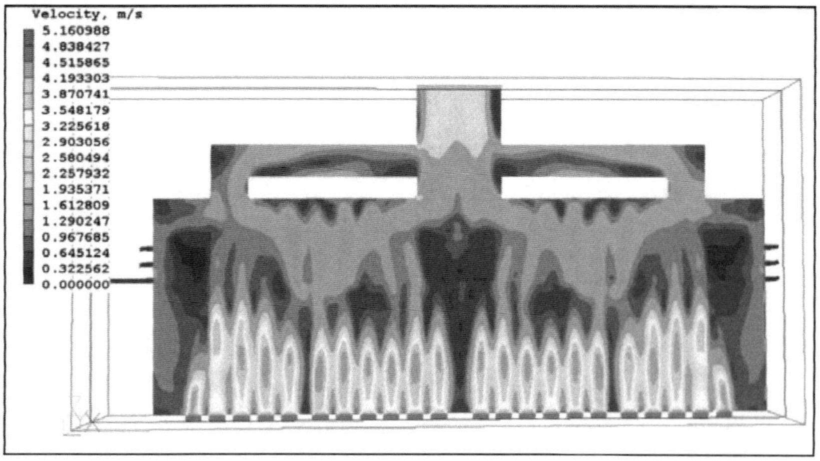

Figure 4 – Velocity profile in the furnace

This result can be compared to the results of the case including radiation and combustion models. The results with all models are sketched in figures below.

Figure 5 shows that the combustion products follow a preferential way. In radiation section, flow is quasi vertical and then it becomes divided in convection section due to the presence of coils.

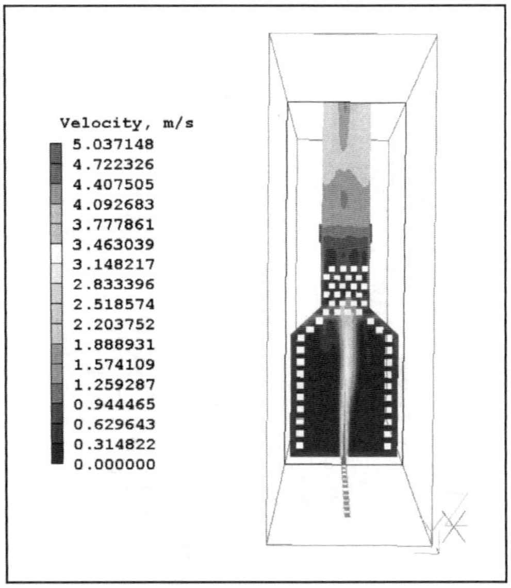

Figure 5 – Velocity profile

Pressure gradient distribution (figure 6) is also important because, in some cases, it can be associated with the velocity by Bernoulli equation. Besides, it can be used as an important tool on structural analysis of furnace identifying where a more resistant material is necessary, for instance.

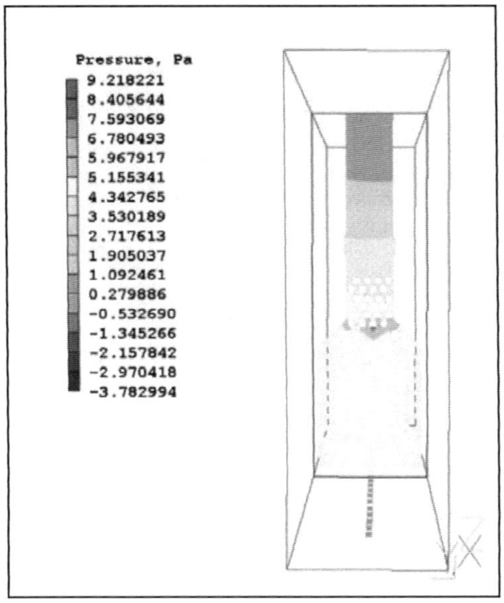

Figure 6 – Pressure profile

Furnace temperature profile (figure 7) helps to understand by what means feed stream is heated in the furnace. This variable profile is quite important for optimization problems, where fuel-air ratio and geometry of the case can be changed so as to achieve desired temperatures. From figure below, is possible to note that convection section (located above the radiation section) is cooler, facilitating recovery of additional heat. On the bottom furnace section main heat transfer mechanism is mainly due to radiation while on the upper section it is due to convection.

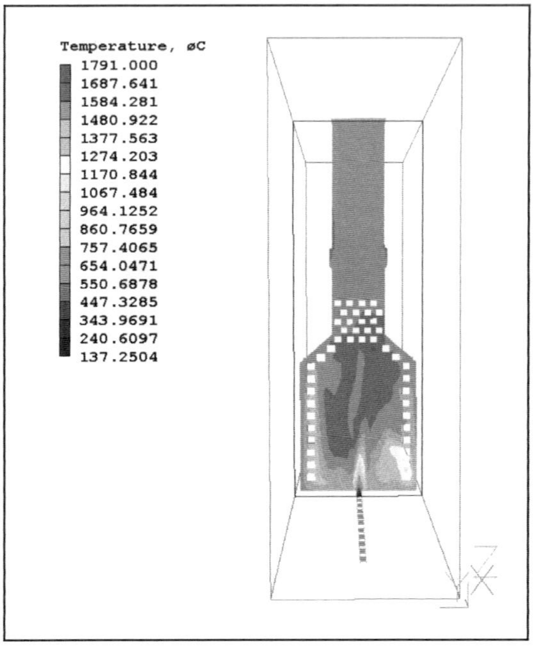

Figure 7 – Temperature profile

Combustion products profile is another considerable tool that allows verifying if flames of the burners are reaching the tubes at convection section. So it can be used in designing a furnace without this undesired condition.

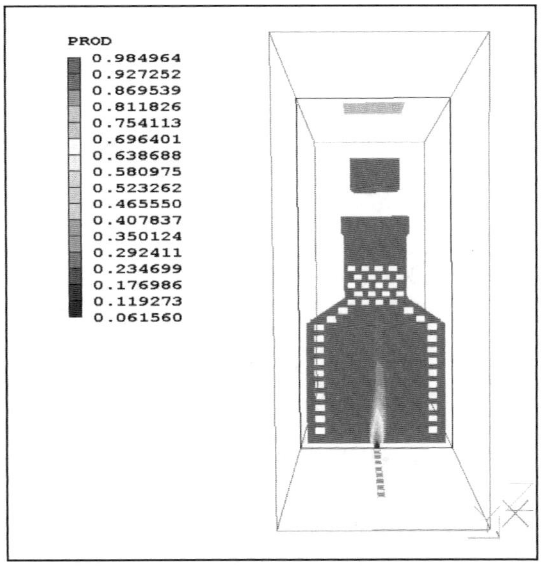

Figure 8 – Combustion products profile

These results are as expected in literature and enable using them for optimization problems for furnace designs.

7) Conclusions

Analysis of the main phenomena that take place inside a furnace was the main objective of this paper. Computational Fluid Dynamics (CFD) techniques have shown that may play an important role in transport phenomena research and it has been used in this paper. PHOENICS (from CHAM) was the commercial software used to work the simulations out. In order to have a complete and coherent result, radiation and combustion models have to be also studied due to their unquestionable importance in a furnace heat transfer mechanisms. IMMERSOL and 3-gases models were used. Simulation results based on these models were consistent as expected.

Furnace design can surely be improved by using CFD tools and it is possible to obtain valuable gains in energy efficiency of this equipment, as observed in results analysis section.

This paper presents only one application of CFD analysis but this tool can be used for a variety of optimization and design problems in process industry, from gas dispersion for sensors positioning to heat transfer in a heat exchanger.

8) References

[1] Anderson, John D., Jr: *Fundamentals of Aerodynamics*, 3rd ed. McGraw-Hill Book Company, Boston, 2001.

[2] PHOENICS Online Encyclopedia.

[3] http://www.csn.com.br, last accessed by September/2007.

[4] http://www.petrobras.com.br, last accessed by September/2007.

[5] LOPES JR., S. Introduction course to heat furnaces. CHEMTECH – A SIEMENS Company.

THERMAL ANALYSIS OF AGRICULTURAL WASTES USED FOR ENERGY PRODUCTION

Erzsébet Cserta, Thermal Processing Technology, University of Leoben, Leoben, Austria

Dr. Harald Raupenstrauch, Thermal Processing Technology, University of Leoben, Leoben, Austria

Dr. Erika Mészáros, Institute of Materials and Environmental Chemistry, Chemical Research Center, Hungarian Academy of Sciences, Budapest, Hungary

Martina Poppenwimmer, Thermal Processing Technology, University of Leoben, Leoben, Austria

Péter Tóvári, Hungarian Institute of Agricultural Engineering, Gödöllő, Hungary

Introduction

The search for new energy sources is a major concern of today's economics and industry. Environmental protection (i.e. lowering harmful emissions, CO_2 neutrality) and decreasing the corrosion of the plants are in the focus of interest at present. However, before an extensive use of a given plant material, a thorough analysis of its composition and thermal behavior is necessary.

The thermal degradation of three different agricultural biomass waste species (triticale grain, corn grain, and cold pressed pumpkinseed residue) was investigated by thermogravimetry/mass spectrometry (TG/MS) and by a 'Differential Scanning Calorimetry' (DSC). Most of all, their burning and pyrolysis behavior is in the focus of interest, so the objective of the present work was to determine the possible applicability interval of these residues. The aim of our work was to study the devolatilization behavior of three samples.

Experiments

Samples

Triticale corn grain and cold-pressed pumpkinseed residue were investigated. The forage crops were obtained from an agricultural plantation in South-Hungary. The oil grain residue was prepared from species grown in the North-Eastern part of Hungary. Prior to the TG/MS analysis, the samples have been air-dried at room temperature and milled using 500 μm sieves. For DSC measurements the samples were not milled because of the easier installation to the burning chamber, and it was also important to pay attention to the similar size of the samples because this parameter considerably influences the results.

Thermo gravimetric / mass spectrometry (TG/MS) analysis

The TG/MS experiments were performed using a Perkin-Elmer TGS-2 thermobalance connected to a Hiden HAL quadrupole mass spectrometer. The measurements were carried out using argon purge gas with a flow rate of 140 ml min^{-1}. The sample size was about 4 mg. The samples were heated from 20°C to 900°C at a rate of 20°C min^{-1}. The mass spectrometer was operated in electron impact ionization mode with 70 eV electron energy.

Differential scanning calorimeter (DSC) analysis

The DSC measurements were performed using a flow calorimeter attached to a thermobalance and a gas analyser. For the tar elimination a tar trap was placed in between the oven and the DSC. The pyrolysis was carried out at 820°C using nitrogen as flush gas. The sample size was about 0.7-1.1g. With this measurement the main characteristics and conditions of the thermal decomposition of the samples were investigated at flash pyrolysis conditions.

Results and discussion

TG/MS analysis

TG/MS is frequently used for the analysis of lignocellulosic materials. This method gives information not only about the devolatilization behavior of a plant material, but also about the composition of the pyrolysed sample [1].

The comparison of the thermogravimetric curves (TG, DTG) of two different forage crops and an oil grain residue are shown in Fig.1. The relative mass loss (TG) and the rate of the decomposition (DTG) depending on temperature are represented in the diagram.

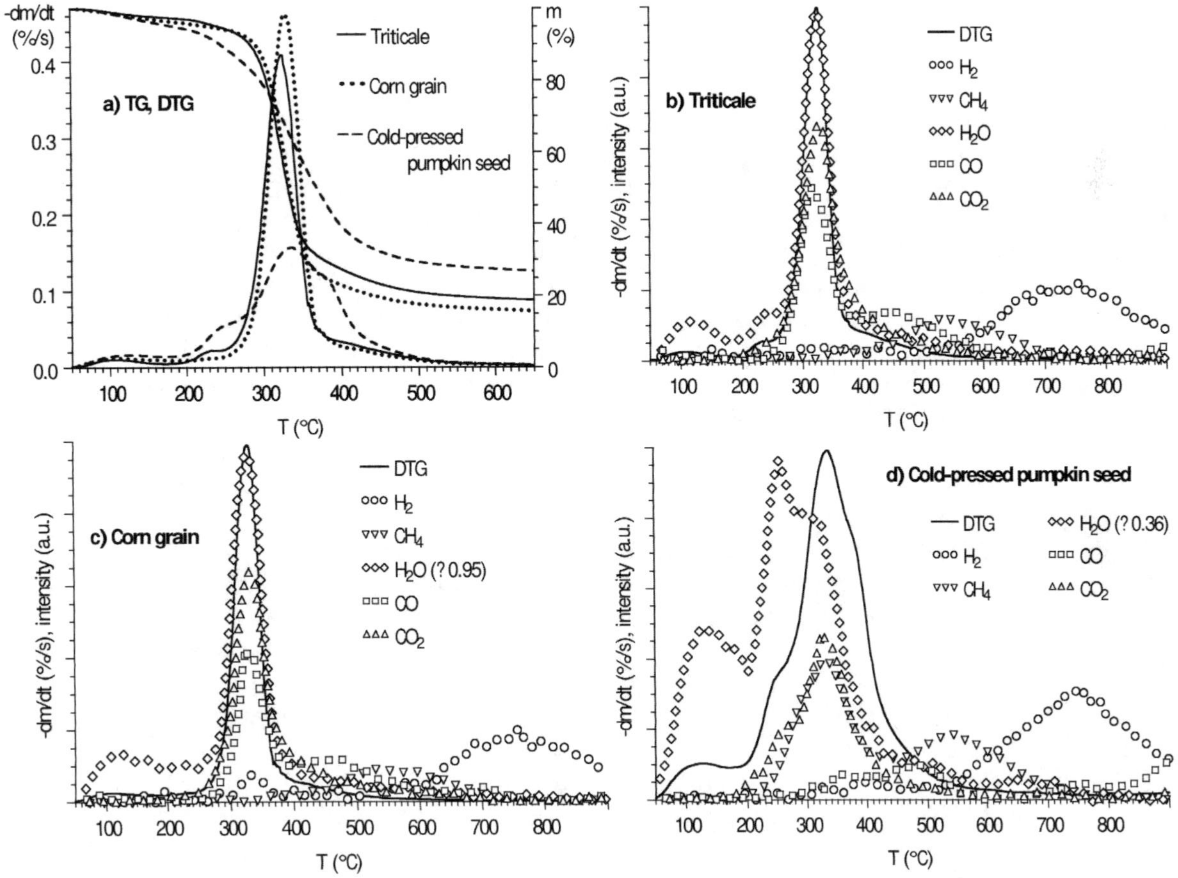

Figure 1. (a)the TG and DTG curves of the samples and the DTG curves and the evalution profiles of some volatiles released from (b) triticale, (c) corngrain and (d) cold-pressed pumpkinseed in argon athmosphere. The DTG curves are magnified to their maxima

The main DTG peak is dominated by the decomposition of cellulose, while the shoulder at lower temperature (at around 270-280 °C) can be attributed mainly to hemicelluloses and extractives decomposition [1,2]. The lignin decomposes at lower rate in a wide temperature range (200-600 °C) [2,3]

Table 1. The thermogravimetric parameters of the samples: DTG peak height (DTG_{max}), peak temperature (T_{peak}), start of hemicelluloses degradation ($T_{hc\ start}$), end of cellulose degradation ($T_{cell\ end}$) and the char yield

Sample	$T_{hc.\ start}$ °C*	$T_{cell.end}$ °C*	T_{pesk} °C	DTG_{max} %/s	Char yield 900 °C/%
Triticale	282	359	324	0.409	18
Corn grain	281	362	332	0.382	16
Pumpkinseed (cold)	-	434	335	0.157	25

As it is suggested by the similarity of the TG and DTG curves, the thermal behavior of triticale and corn grain are rather similar (Fig. 1a). The decomposition of corn grain takes place at a somewhat higher rate, yielding a smaller amount of char compared to the triticale sample. The great similarity in the decomposition can be explained by the similar chemical composition of the samples [1-3,6]. A possible reason for the difference in the rate of decomposition is the different amount of inorganic ions in the two forage crops. As it is well-known [1,3-6] the inorganic ions have a catalytic effect in the thermal decomposition of natural polymers, which make up the lignocellulosic materials. The larger the amount of inorganic compounds, the larger the carbonaceous residue and the smaller the rate of decomposition are.

The decomposition of the cold-pressed pumpkin seed proceeds in about the same temperature interval as that of the forage crops (Fig. 1.a.) (i.e. between about 200–500 °C), however, the thermal behavior of this sample is utterly different. The carbonaceous residue is significantly larger than in the case of the other two samples. The DTG peak is considerably smaller implying that the maximum rate of decomposition is lower. On the other hand, two shoulders are discernible on the main DTG peak. The shoulder at about 280 °C can be attributed to the decomposition of hemicellulose, lignin and extractive compounds, whereas the shoulder around 380 °C corresponds probably to extractives and lignin decomposition [2-4].

The mass spectrometric (MS) curves give information about the evolution of volatile compounds. As an example, the formation of hydrogen, methane, water, carbon monoxide and carbon dioxide are shown in Figs. 1.b-d. The decomposition of all samples starts with the elimination of adsorptive water (50 – 200 °C)[2]. In the main decomposition process water, carbon monoxide and carbon dioxide are released. Methane formation starts above about 400 °C, whereas hydrogen evolves above 550 – 600 °C. The double peaked curve of methane in case of the pumpkinseed (Fig. 1.b.) corresponds to two different decomposition processes, but methane has only one peak in the case of the forage crops (Fig. 1.b,c.). Further volatile compounds were also studied, but these results are not published in this articel.

The comparison of the DTG curves of the forage crops and the pumpkin seed suggests that

- The thermal behavior of triticale and corn grain are rather similar whereas that of the pumpkinseed residue are different
- The decomposition of the pumpkin seed proceeds in about the same temperature interval as that of the forage crops
- A significantly smaller amount of water is released from the pumpkin seed sample, whereas the yields of the other products are comparable.
- The pumpkin seed sample contains a larger amount of extractives and lignin than triticale and corn grain

* The start of the hemicellulose decomposition (onset temperature) and the end of the cellulose decomposition (offset temperature) have been determined by extrapolation of the DTG curves.

DSC analysis

In the next part of the present work volatiles (including tars) of biomass and tars from biomass were investigated with flash pyrolysis. The calorific value, the relative mass loss, reaction heat and oxygen demand of the samples will be described.

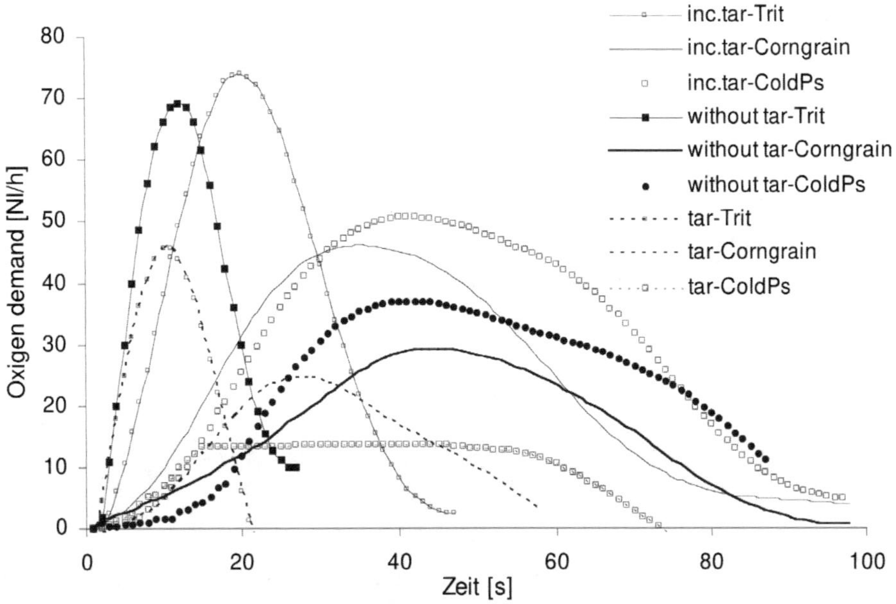

Figure 3. Oxygen demand of reaction of tar, average of all volatiles and average of non-condensable fraction of triticale grain, cold pressed pumpkinseed and corn grain under flash pyrolysis conditions (T_{oven} = 825°C, nitrogen atm, $T_{catalyst,start}$ =510°C)

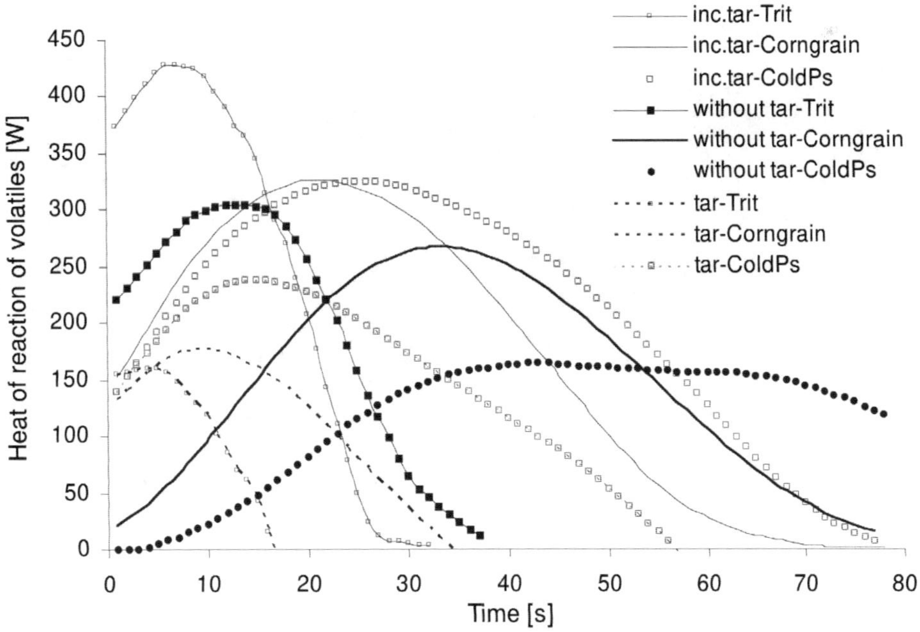

Figure 4. Heat of reaction of tar, average of all volatiles and average of non-condensable fraction of triticale grain, cold pressed pumpkinseed and corn grain under flash pyrolysis conditions (T_{oven} = 825°C, nitrogen atm, $T_{catalyst,start}$ =510°C)

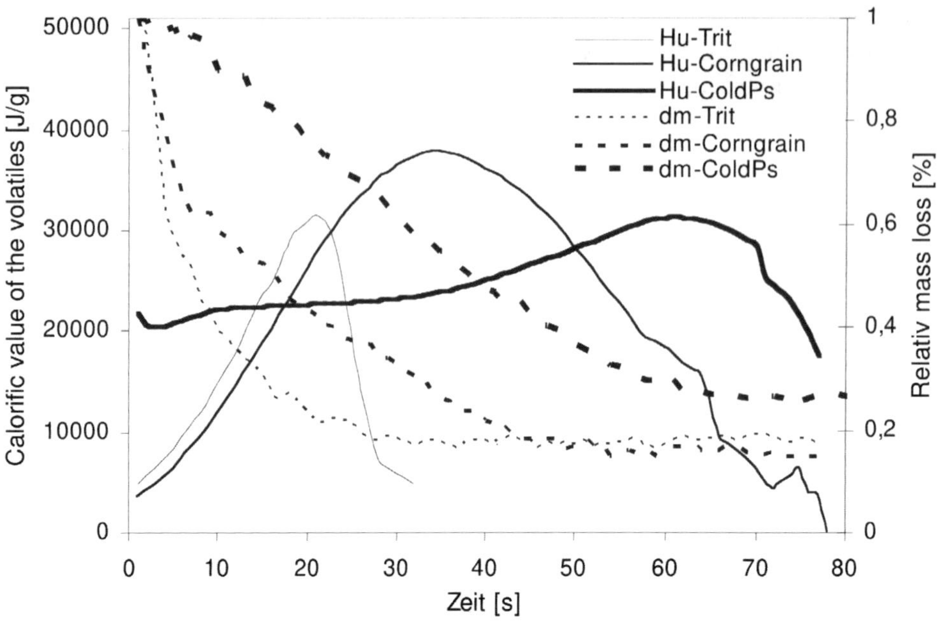

Figure 5. Calorific value and mass loss dependent on time of triticale grain, cold pressed pumpkinseed and corn grain under flash pyrolysis conditions (T_{oven} = 825°C, nitrogen atm, $T_{catalyst,start}$ =510°C)

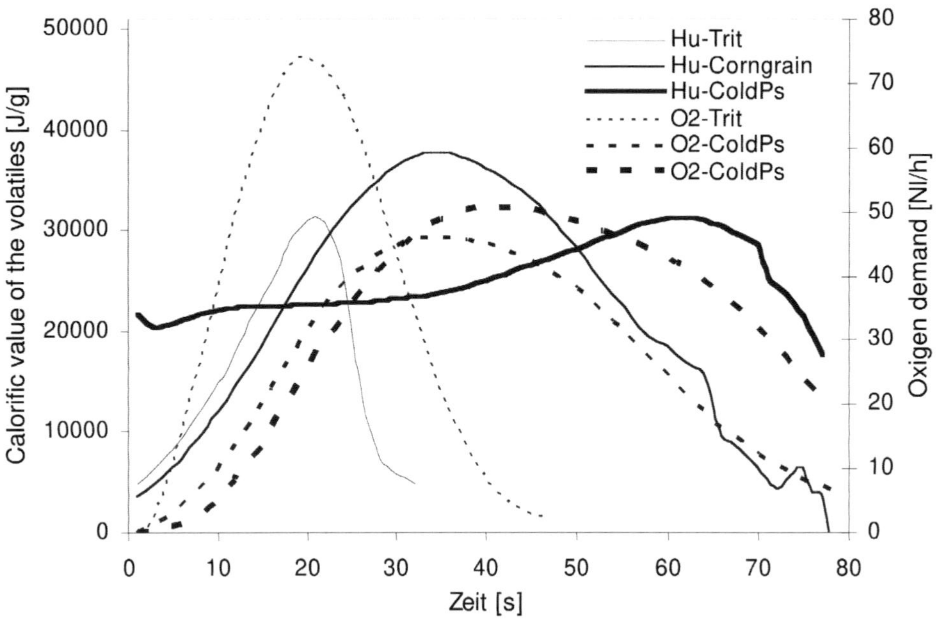

Figure 6. Calorific value and oxygen demand dependent on time of triticale grain, cold pressed pumpkinseed and corn grain under flash pyrolysis conditions (T_{oven} = 825°C, nitrogen atm, $T_{catalyst,start}$ =510°C)

The oxygen demand of volatiles means the total amount of O_2 required for the whole process. It shows a similar behavior as the heat of reaction, the curve rises to a maximum at the beginning. After the maximum is reached the oxygen demand decreases towards zero until the end of the pyrolysis is reached.

At the end of the pyrolysis when no more volatiles form the curve of the relative mass loss gets linear. The curves of the calorific value of triticale and corn grain have no peak at the end of the decomposition process. This peak refers to the rapid heating and the great mass loss because a big reaction surface exists at the beginning of pyrolysis. In case of pumpkinseed contrary to the others the curve of calorific value shows another peak at the end of the pyrolysis. This second one denotes the secondary crack reactions which appear when the formed pyrolysis gas passes through the already formed ash and carbon frame work. This second maximum can be assigned to a small amount of hydrogen rich gas with a high specific calorific value because in the same time no shoulder can be observed on the curve of the oxygen demand of pumpkinseed [7-10].

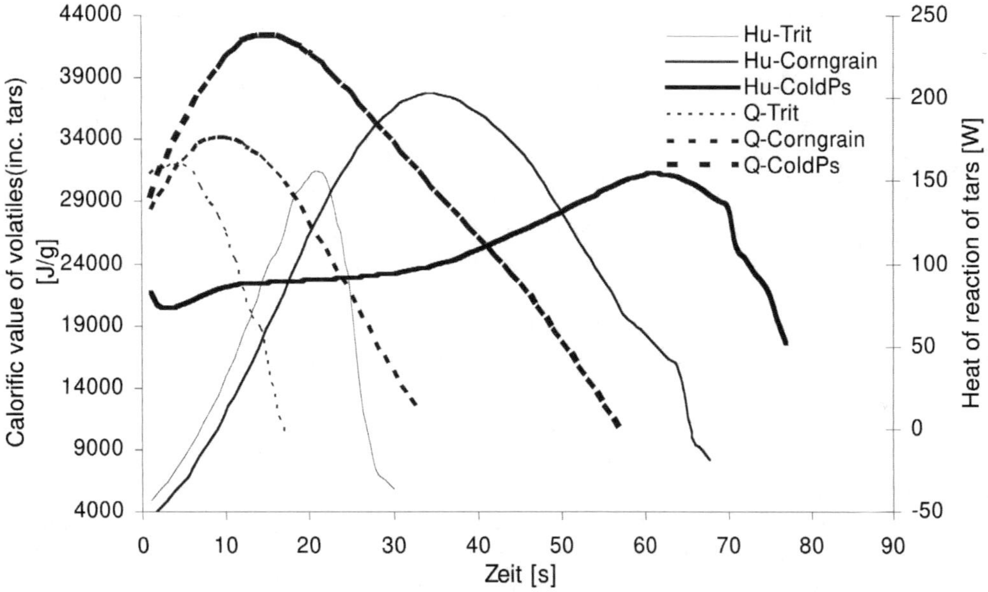

Figure 7. Calorific value of volatiles (inc. tars) and heat of reaction of tars dependent on time analysing triticale, corn grain and cold pressed pumpkinseed under flash pyrolysis conditions (T_{oven} = 825°C, nitrogen atm, $T_{catalyst,start}$ =510°C)

The tar is generated especially in the first half of the pyrolysis. It runs to approximately the 30% of the heat of reaction. At the very end of the thermal decomposition no influence of tars on the heat of reaction of the volatiles can be detected. In Figure 5 one can see that when the peak at the end of the curve of the calorific value of all volatiles reaches the maximum, the heat of reaction of the volatiles of tars equals zero. Hence the formed volatiles after that last decrease of the heat of reaction of the tars are non-condensable volatiles. This means that the higher molecular hydrocarbons formed at the end of the pyrolysis time are cracked on their way from the centre of the particle through the already formed ash and carbon framework to the outer surface [8-11].

We have discussed first of all about the emission problems because of the high volatile content of the solid fuels. The char yield varies between 15 and 26 w% which refers to an average 75-85 w% of volatiles suited for pyrolysis. This amount of useful gases proves the

importance of these biofuels. At higher heating rate (from ~ 1K/s till ~1.000 K/s) a significant amount of tar is formed because the higher beginning temperature of decomposition by faster heating, where the temperature of pyrolysis coke decreases rapidly and so tar can not crack on the surface of samples[9].
From the above investigation the next summary can be given:

- ∞ Although tar itself is an undesirable product of flash pyrolysis, its energy content is worth to be used after catalytic cracking. A significant share of the heat of reaction of the volatiles is supplied by tars in the first half of the pyrolysis process.
- ∞ With the help of the applied method it was possible to determine the heat of reaction and the oxygen demand of the samples with and without tar furthermore the calorific value and the relative mass loss of the volatiles including tar.
- ∞ The remarkable variances in the oxygen demand denote the differences in the pyrolysis process. The reason for these differences needs to be looked at for different chemical compositions, densities and geometries of the defined materials.

Conclusions

The coherences between the thermochemical behaviors of the samples can be established from the research presented above:

Triticale burns out fast and has a really low char yield. This big volatile content gives a good efficiency and a high energy yield by pyrolysing it, furthermore the heat content of tar plays a minor role of the total reaction heat as in case of the other examined organic materials. This crop seems to be rather adequate for energy generation.

Corn grain could be a suitable biofuel too, because of its high volatile yield, longer decomposition time with a relative high calorific value and the average tar formation at the beginning of the pyrolysis. The chemical composition and consequently the thermal behavior of triticale and corn grain are rather similar, whereas that of the pumpkinseed residue are by far different.

Pumpkinseed residues have big char contents at the end of the pyrolysis. This fact relates to the high inorganic content of the sample, which has a strong influence on the temperature of pyrolysis as well as on the product distribution. It has a slower and longer decomposition process which can be caused by the relative low amount of volatile. At flash pyrolysis a secondary pyrolysis occurs, where a small amount of hydrogen rich gas with a high specific calorific value were produced.

Aknowledgements
This work was supported by Graz University of Technology and the Hungarian National Research Fund (OTKA K61504, T037705).

References

1. M. Müller-Hagedorn, H. Bockhorn, L. Krebs, U. Müller (2003), "A comparative kinetic study on the pyrolysis of three different wood species," *J. Anal. Appl. Pyrolysis*, 68-69, pp. 231-249.
2. E. Mészáros, E. Jakab, G. Várhegyi, P. Szepesváry, B Marosvölgyi (2004), "Comparative study of thermal behavior of wood and bark of young shoots obtained from an energy plantation," *J. Anal. Appl. Pyrolysis*, 72, pp.317-328.
3. E. Mészáros, G. Várhegyi, E. Jakab (2004), "Thermogravimetric and Reaction Kinetic Analysis of Biomass Samples from an Energy Plantation," *Energy&Fuel*, 18, 497-507.

4. J. Fengl, Q. YuHong, A. E. S. Green (2006), "Analytical model of corn cob Pyroprobe-FTIR data," *Biomass and Bioenergy,* 30, pp. 486–492
5. L.O. Pordesimo, B.R. Hames, S. Sokhansanj, W.C. Edens (2005), "Variation in corn stover composition and energy content with crop maturity," *Biomass and Bioenergy,* 28, pp. 366–374
6. A. Heinz, M. Kaltschmitt, R. Stülpnagel, K. Scheffer (2001), "Comparison of moist vs. air-dry biomass provision chains for energy generation from annual crops," *Biomass and Bioenergy*, 20, pp. 197–215
7. H. Raupenstrauch, C. Maurer, P. Zottler (2006), "Measurements of the Calorific Value of Volatiles and Oxygen Demand of Solid Fuels as a Function of Time with a Mobile Analyser," in *The 2006 Annual Meeting-AIChE Conference,* ed. San Francisco, CA,
8. M. Poppemwimmer (2007), "Reaktionswärmestrom von Flüchtigen und Teer in Abhängigkeit der Zeit bei der Pyrolyse von festen Brennstoffen, " *Diplomwork,* Technische Universität, Graz
9. C. Maurer (2006), "Development of a Differential Scanning Calorimeter - Experimental Investigations of Wood and Plastics," *Dissertation*, Technische Universität, Graz
10. A.R. Fraga, A. F. Gaines, R. Kandiyoti (1990), "Characterization of biomass pyrolysis tars produced in the relative absence of extraparticle secondary reactions" *Fuel,* July, pp.803-809.
11. F. J. Mastral, E. Esperanza, C. Berrueco, M. Juste, J. Ceamanos (2003), "Fluidized bed thermal degradation products of HDPE in an inert atmosphere and in air–nitrogen mixtures," *J. Anal. Appl. Pyrolysis*, Oct, pp.1-17.

AUTHOR INDEX

AUTHOR INDEX